**Oryx Sourcebook Series
in Business and Management**

Chemical Industries
An Information Sourcebook

Oryx Sourcebook Series in Business and Management
Paul Wasserman, Series Editor

1. Small Business
2. Government Regulation of Business
3. Doing Business in and with Latin America
4. Decision Making
5. The Real Estate Industry
6. Public Finance
7. Women in Administration and Management
8. Retirement Benefit Plans
9. Chemical Industries

**Oryx Sourcebook Series
in Business and Management**

Chemical Industries
An Information Sourcebook

by Phae H. Dorman

Manager, Business Information Center,
The Dow Chemical Company

Phoenix • New York
ORYX PRESS
1988

The rare Arabian Oryx is believed to have inspired the myth of the unicorn. This desert antelope became virtually extinct in the early 1960s. At that time several groups of international conservationists arranged to have 9 animals sent to the Phoenix Zoo to be the nucleus of a captive breeding herd. Today the Oryx population is over 400 and herds have been returned to reserves in Israel, Jordan, and Oman.

Copyright © 1988 by
The Oryx Press
2214 North Central at Encanto
Phoenix, Arizona 85004-1483
Published simultaneously in Canada

All rights reserved
No part of this publication may be reproduced or transmitted in any form or by any means, electronic or mechanical, including photocopying, recording, or by any information storage or retrieval system, without permission in writing from The Oryx Press.

Printed and Bound in the United States of America

The paper used in this publication meets the minimum requirements of American National Standard for Information Science—Permanence of Paper for Printed Library Materials, ANSI Z39.48, 1984.

Library of Congress Cataloging-in-Publication Data

Dorman, Phae H.
  Chemical industries : an information sourcebook / by Phae H. Dorman.
  p. cm. — (Oryx sourcebook series in business and management; no. 9)
  Includes indexes.
  ISBN 0-89774-257-5
  1. Chemical industry—Bibliography. I. Title. II. Series.
Z5521.D65 1988
[HD9650.5]
016.3384′766–dc19   87-23180

Contents

Introduction   vii
   Scope of the Study   vii
   Standard Industrial Classification System   viii

Core Library Collection   1
   Bibliographies   1
   Biographical Directories   2
   Electronic Retrieval Systems   3
   Encyclopedias, Dictionaries, and Handbooks   7
   Indexes, Abstracts, and Continuing Services   12
   Newspapers, Journals, and Newsletters   16
   U.S. Government Publications   21

Categorical Arrangement   24
   Agricultural Chemicals, Food and Beverages   24
   Cleaning Preparations, Cosmetics and Toiletries   30
   Drugs, Pharmaceuticals   31
   Electronics, Electrical   35
   Mining, Minerals, Metals   39
   Paints, Varnishes, Lacquers, Enamels, Coatings, Adhesives and Sealants   45
   Paper and Allied Products   47
   Petrochemicals, Energy   50
   Plastics Materials, Packaging   56
   Rubber   59
   Textiles   61

Associations   64
   General   64
   Agricultural Chemicals, Food and Beverages   65
   Cleaning Preparations, Cosmetics and Toiletries   66
   Drugs, Pharmaceuticals   66
   Electronics, Electrical   67
   Mining, Minerals, Metals   68
   Paints, Varnishes, Lacquers, Enamels, Coatings, Adhesives and Sealants   69
   Paper and Allied Products   69

Petrochemicals, Energy 70
Plastics Materials, Packaging 70
Rubber 71
Textiles 72

Consultant Sources 73
Directories 73
General 74
Agricultural Chemicals, Food and Beverages 76
Cleaning Preparations, Cosmetics and Toiletries 76
Drugs, Pharmaceuticals 76
Electronics, Electrical 77
Mining, Minerals, Metals 77
Paints, Varnishes, Lacquers, Enamels, Coatings, Adhesives and Sealants 78
Paper and Allied Products 78
Petrochemicals, Energy 78
Plastic Materials, Packaging 79
Textiles 79

Author Index 81

Title Index 83

Subject Index 92

Introduction

Contrary to the impressions created by news media coverage of the chemical industry, chemicals are not bad for humankind. In fact, they may be crucial to life; they can prolong life and even add to the quality of life.

The chemical industry, with its many manifestations, is woven into every aspect of economic life and is thus very difficult to define. However, for purposes of clarity, it may be helpful to provide a brief description. This segment of the economy "encompasses inorganic and organic chemicals employed in industry, plastics, drugs and other biomedical products, rubber, fertilizers and explosives, and on and on."*

The primary focus in this study is the chemical process industry, which is sometimes referred to as the CPI or chemicals and allied products industry. It includes industries involved in the manufacture of organic and inorganic chemicals, petroleum, petrochemicals, rubber, plastics, fertilizers or agrichemicals, paint, paper, synthetic fibers, pharmaceuticals, food additives, and flavors and dyes.

The selection of materials in this bibliography will focus on those items determined to be most useful in providing chemically related business information of a factual nature, which can be used to conduct comprehensive analyses of chemical markets.

SCOPE OF THE STUDY

In order to manage their businesses efficiently and profitably, workers in the chemical and allied products industries have a great need for a large variety of current business-related information. Knowing what is available and where and how to find it can be a frustrating and time consuming experience. This collection of sources is intended as a guide to seekers of such information. Anyone wanting business-related information about the chemical industry could be asking for anything from a biography of a scientist to a phone

---

*George C. Pimental, ed. *Opportunities in Chemistry* (Washington, DC: National Research Council, National Academy Press, 1985), p. 209.

number of an official in the U.S. Environmental Protection Agency. Most often, though, the kinds of questions asked are concerned with production, producers, capacities, prices, process economics, and end uses. In one word, what is the market? Also, suppliers of chemicals to other industries are interested and involved in what goes on in those industries. Therefore, in an attempt to provide a useful arrangement of listings and descriptions of sources, grouping of the materials will be based partly on the categories in the Standard Industrial Classification System and partly on an analysis done by the Charles H. Kline & Co., Inc., which identifies seventeen of the leading industries served by chemical distributors. This analysis was based on the combined sales of the 126 leading chemical distributors in 1985. This categorical arrangement of sources will follow the Core Library Collection. In all the categories, including the Core Library Collection, the materials will be subdivided by specific types of reference materials such as biographical directories, electronic retrieval systems, etc.

An attempt also has been made to provide the most current sources available and to concentrate on national sources or those available in the English language. This reference guide will also include listings of other sources of chemical information such as associations and consultants and an annotated bibliography of recent books on chemical information sources.

STANDARD INDUSTRIAL CLASSIFICATION SYSTEM

The Standard Industrial Classification System (SIC) is a numbering system developed by the U.S. Department of Commerce to facilitate compilation and presentation of data for uniformity and comparability. *The Standard Industrial Classification Manual* (1972) and *Supplement* (1977), published by the Office of Management and Budget, gives the SIC numbers for manufacturing and nonmanufacturing establishments and also defines each category. SIC codes are revised periodically to reflect changes in the industrial makeup of the economy. The Office of Management and Budget published a notice in the *Federal Register*, February 22, 1984, of an intent to revise the SIC for 1987. The Technical Committee on Industrial Classification was established and has been evaluating changes submitted by businesses, trade associations, individuals, and government agencies. The revision became effective January 1, 1987.

Of the changed classifications, two are of significant interest to the chemical industry. The computer and computer-peripherals industry will gain fifteen new listings. The Miscellaneous Plastic Products category (3079), will be broken down into nine new categories. The SIC system is used by most federal statistical agencies, most state agencies, and many private organizations to classify all manufacturing and nonmanufacturing industries. Chemical industry management uses the SIC system in economic and sales forecasting, marketing analysis, and so on. It is employed as a market research tool by many businesses in the classification of their customers and suppliers. By using SIC numbers or codes, information on nearly every corporation and business activity can be compiled and compared. But before you can use this information, you must first determine the SIC number. The following steps describe how to determine the SIC number:

- The first two digits of the SIC code indicate the major group:
  Ex: 28 is the major group for chemicals and allied products.
- The first three digits of the code represent the subgroup within a major group:
  Ex: 283 is the subgroup number for drugs.
- The four-digit number defines the specific industry within a subgroup:
  Ex: 2831 represents a chemical or allied product that is a drug in the biological products industry.

Finding SIC Numbers

By developing an SIC code number for any specific type of industry, you can then use a number of sources to identify specific companies or potential customers or accounts, or determine size, growth patterns, or potential of an industry. For example, if a chemical company wanted to make an evaluation of a new chemical product, the company would first determine what industries of products it would be used in and then define the industry or product by an SIC number. This number can then be used in various sources such as the *Census of Manufactures* to determine how large the industry is and in the *U.S. Industrial Outlook* to make an estimate of the growth potential. Other directories, reference books, and online databases provide information using the standard industrial classification as a format.

Chemical Industries

Core Library Collection

BIBLIOGRAPHIES

1. Antony, Arthur. *Guide to Basic Information Sources in Chemistry.* Information Resources Series. New York: John Wiley & Sons, Inc., 1979. 219 p.
>Aimed at students of chemistry. Lists general sources of data, a section on searching by computer, search strategies, and other standard references on chemical data.

2. Grayson, M. *Information Retrieval in Chemistry and Chemical Patent Law.* New York: John Wiley & Sons, Inc., 1983. 116 p.

3. Maizell, R. E. *How to Find Chemical Information; A Guide for Practicing Chemists, Teachers & Students.* 2d ed. New York: John Wiley & Sons, Inc., 1986. 496 p.

4. Peck, Theodore P. *Chemical Industries Information Sources.* Management Information Guide Series, vol. 29. Detroit, MI: Gale Research Co., 1979. 595 p.
>Lists many sources of data used in chemical and related industries and focuses on subject areas of interest to chemical engineers.

5. Sturchio, Jeffrey L., ed. *Corporate History and the Chemical Industries; A Resource Guide.* Publication No. 4. Philadelphia, PA: Center for the History of Chemistry, 1985. 53 p.
>Contains an annotated bibliography on general works, chemistry and industry in America, industrial research, and company histories.

6. Wolman, Yecheskel. *Chemical Information: A Practical Guide to Utilization.* New York: John Wiley & Sons, Inc., 1983. 191 p.
>The individual chemist or scientist is the audience for this guide to the chemical literature. Some of the information presents details on how to locate data, how to conduct a literature search, online searching techniques, patent information, and a short section on chemical marketing and process information.

General Business Information

7. Daniells, Lorna M. *Business Information Sources.* Rev. ed. Berkeley and Los Angeles, CA: University of California Press, 1985. 673 p.
   A guide to general business reference sources including annotations and practical information on using the materials.

BIOGRAPHICAL DIRECTORIES

These resources yield information about notable persons and specialists in the fields of science, business, and industry. An interesting use of the online versions of many of these data sources is the retrieval of listings of specialists in a specific field along with lists of their publications or contributions.

8. *American Men and Women of Science; Physical and Biological Sciences.* 16th ed. New York: R. R. Bowker Co., 1986. 7 vols.
   Biographical data on U.S. and Canadian scientists in the physical and biological sciences. Records include name, personal statistics, discipline, education, honorary degrees, professional experience, positions, honors and awards, memberships, research, addresses. Available also on BRS and DIALOG.

9. *Chemical Engineering Faculties, 1986-87.* Vol. 35. New York: American Institute of Chemical Engineers, 1986. 256 p. (Annual)
   Covers faculty members in chemical engineering departments in schools in the U.S. and in foreign schools.

10. *Marquis Who's Who in America, 1986-87.* 44th ed. Chicago: Marquis Who's Who, Inc., Macmillan Directory Division, 1986.
   A comprehensive biographical source on leading and influential business executives and scientists.

11. NATIONAL REGISTRY OF WOMEN IN SCIENCE AND ENGINEERING. Washington, DC: Association for Women in Science Registry.
   A database including the names and qualifications of female scientists and engineers.

12. *Profiles of Eminent American Chemists.* Arvada, CO: Litarvan Literature, 1987. 600 p.
   Biographical information on more than 100 winners of the American Institute of Chemists Gold Medal and Chemical Pioneer Awards. (Not published as of fall 1987.)

13. *Standard & Poor's Register of Corporations, Directors and Executives.* Vol. 2. New York: Standard & Poor's, 1986.
   Lists individuals serving as officers, directors, trustees, partners, etc., in business and professional organizations. Information listed includes year and place of birth, college graduation, and fraternal memberships.

14. *Who's Who in Finance and Industry, 25th ed.* Wilmette, IL: Marquis Who's Who, Inc., Macmillan Directory Division, 1987. 775 p.
> Biographical profiles of business professionals from a wide variety of fields. Provides information on occupation, education, family, publications, awards, religion, and addresses.

15. *Who's Who in Frontiers of Science & Technology.* 2d ed. Chicago: Marquis Who's Who, Inc., Macmillan Directory Division, 1985.
> Provides professional and biographical information on leaders in current fields in science and technology.

16. *Who's Who in Technology.* 5th ed. Woodbridge, CT: RP Research Publications, 1986. 7 vols.
> Formerly *Who's Who in Technology Today*, this set contains biographies of top scientists, technology experts, and engineers arranged by discipline.

ELECTRONIC RETRIEVAL SYSTEMS

According to Quadra Associates, publishers of the *Directory of Online Databases*, the total number of available databases in 1987 is over 3,300, compared to 400 in 1979. Also, online databases may be bibliographic, numeric, or full text and offered for access on a specific system with other databases. Listed here are some major systems or providers of databases for online access.

17. BRS/SEARCH SERVICE. BRS Information Technologies, 1200 Route 7, Latham, NY 12110.
> This system provides access to about ninety databases in the areas of the life sciences, medicine, biosciences, engineering, business, economics, and others. Files of particular interest include:
> ACS Directory of Graduate Research
> ACS Journals Online
> Association's Publications in Print—pamphlets, journals, newsletters and other printed materials published by trade associations in the U.S. and Canada.
> Corporate and Industry Research Reports Online
> Health Planning and Administration
> IHS Vendor Information
> Kirk-Othmer Encyclopedia of Chemical Technology
> Index to Frost & Sullivan Market Research Reports

18. CEH ON-LINE. SDC/ORBIT-SRI International, 8000 Westpark Dr., McLean, VA 22102.
> This is an online service consisting of the data available in the *Chemical Economics Handbook Program*. The reports and product reviews are available in full text through System Development Corporation's ORBIT Information Retrieval System.

19. DATA RESOURCES, INC. 24 Hartwell Ave., Lexington, MA. 02173.

DRI's computer-accessible data provides information from national, international, regional, financial, industrial, company, and special purpose data series. Many of these are historical in nature and contents include:
- U.S. national and regional databanks
- U.S. census, county patterns, etc.
- International databanks
- IMF and OECD publications
- Industrial/sectoral databanks
- Chemical, energy, fertilizer, pulp and paper, etc.

The chemical databases contain information on economic, market, and financial conditions relating to the chemical industry. This includes the following:

General indicators by type of chemical:
- Production
- Sales
- Prices
- Foreign Trade

Plant capacities by:
- Product
- Process
- Company
- Location

Coverage:
- U.S., company, plant, chemical product
- Annual, quarterly, and monthly
- 9,000 series

20. DIALOG INFORMATION RETRIEVAL SERVICES, Inc. DIALOG Information Services, Inc., 3460 Hillview Ave., Palo Alto, CA 94304

An electronic system that has over 250 different databases which cover a wide variety of subjects. There are some files that are specific to chemical business information, and DIALOG has just added a new DIALINDEX category which includes files containing chemical business news. These files contain information on company mergers and acquisitions, sales and marketing activities, personnel news, and many other items. The following files are included in the category CHEMBUS:

File 16—PTS PROMT
Files 18, 98—PTS F & S Indexes
File 19—Chemical Industry Notes
File 111—National Newspaper Index
Files 132, 134—Standard & Poor's News
File 148—Trade & Industry Index
File 211—Newsearch
File 319—Chemical Business Newsbase
File 545—INVESTEXT

Other very useful files are:
File 269—Materials Business File
Files 500-511—Electronic Yellow Pages
Files 516-519—Dun & Bradstreet Files
Files 531-532—Trinet Databases

File 535—Thomas Register Online
File 536—Thomas New Products
Files 555-557—Moody's Files
File 570—PTS New Product Announcements
File 580—CENDATA—U.S. Department of Commerce, Bureau of the Census. Includes current industrial reports on inorganic chemicals, inorganic fertilizer materials and related products and paint, varnish, and lacquer.
File 602—Business and Industry

21. DOW JONES NEWS RETRIEVAL. Dow Jones & Co., Inc., P.O. Box 300, Princeton, NJ 08540.
Stories in full-text format from the *Wall Street Journal, Barron's*, and the *Dow Jones Newswire*. Also contains stock quotes, the *Academic American Encyclopedia*, weather reports, and other information.

22. The Economic Bulletin Board. Office of Business Analysis, Main Commerce Bldg. Washington, DC 20230: U.S. Department of Commerce.
A source for current economic information from news releases of governmental agencies. Kinds of data available include gross national product, the employment situation, personal income, consumer price index, producer price index, and economic indicators.

23. ELSS—Electronic Legislative Search System. Commerce Clearing House, 4025 W. Peterson Ave., Chicago, IL 60646.
Information on legislative proposals in the form of bills and resolutions in all fifty states. Database is updated daily.

24. INVESTEXT. Business Research Corp., 12 Farnsworth St., Boston, MA 02210.
A full-text database of company and industry research compiled by industry specialists at major Wall Street and regional brokerage houses and financial research firms worldwide.

25. NERAC. NERAC, Inc., 1 Technology Dr., Tolland, CT 06084.
Nonconfidential research developed in NASA's laboratories and space missions plus 105 additional databases that cover over 30,000 published journals. The service was established to provide American industry access to technical and business information drawn from NASA and other government agencies as well as professional and academic organizations worldwide. Partial list of databases accessed by NERAC include:
ABI Inform
Agricultural Abstracts
Biological Research Index
CA Search
Ceramics Abstracts
Coal Abstracts
Electrical and Electronics Abstracts
Food Science & Technology Abstracts
Frost & Sullivan Market Reports
Marketing and Advertising Reference Service
Materials Business File

Metals Abstracts
New Product Announcements
Packaging Science and Technology Abstracts
Paper and Board Industries
PTS PROMT
Rubber and Plastics Research Abstracts
World Aluminum Abstracts
World Surface Coatings Abstracts
World Textiles Abstracts

26. NEWSNET. NewsNet, Inc., 945 Haverford Dr., Bryn Mawr, PA 19010.
Available on this service are more than 300 newsletters, wire services, and other publications in a full-text format.

27. NEXIS. Mead Data Central, Inc., 9393 Springboro Pike, P.O. Box 933, Dayton, OH 45401.
Contains the full text of general and business news and other information from newspapers, magazines, wire services, newsletters, and reference sources.

28. PERGAMON INFOLINE. Pergamon Infoline, Inc., 8000 Westpark Dr., McLean, VA 22102.
These databases contain information on patents, manufacturing technology, health and safety, biosciences, construction and engineering, mining and minerals, business, and law. Files that include chemical information are:
Chemical Age Project File—A detailed list of over 14,000 chemical and process plant projects; updated daily.
Chemical Business Newsbase—London: The Royal Society of Chemistry. Covers trends and current affairs in the chemical industry and its end markets.
Fine Chemicals Directory—Fraser Williams Scientific Systems, Ltd. Lists commercially available research chemicals and where they can be obtained.
Rapra Abstracts—Produced by the Rubber and Plastics Research Association, this file provides information on all aspects of the trade. Material is abstracted from international periodicals, conference proceedings, trade and technical literature, books, and standards.
Standard Industrial Classification, United Kingdom Central Statistical Office—A database of United Kingdom SIC codes and headings.
U.S. Standard Industrial Classification, U.S. Department of Commerce—A database of U.S. SIC codes corresponding to the 1972 published manual plus the 1977 supplement.

29. PIERS. Port Import Export Reporting Service, The Journal of Commerce, 110 Wall St., New York, NY 10005.
This service from the *Journal of Commerce* contains information on all U.S. waterborne cargoes to and from foreign countries. The information is compiled by PIERS reporters in fifty major U.S. ports who transcribe all manifests of vessels entering or departing those ports. Records for each cargo item include the name of the importer or exporter, weight

and/or amount of item, name of ship, port of loading or discharge, and name of shipper or consignee.

30. SDC/ORBIT Search Service. 8000 Westpark Dr., McLean, VA 22102.

The ORBIT (Online Retrieval of Bibliographic Information Timeshared) databases cover a wide variety of subjects and also have some industry specific databases. Of particular interest are the following:
ASI—American Statistics Index
CEH ONLINE—The Chemical Economics Handbook, Online SRI, International
CIN—Chemical Industry Notes, ACS
CIS—Congressional Information Service
EDB—U.S. Department of Energy
LABORDOC—International Labour Organization.

31. TRADSTAT. Data Star Marketing Ltd., Plaza Suite, 114 Jermyn St., London SW1Y 6HJ England.

Import and export data is available for about ten countries. This is a source for data that does not involve the U.S. as an exporter or importer. Production information can also be accessed for about 250 chemical commodities.

32. VU-TEXT. VU/Text Information Services, Inc., 1211 Chestnut St., Philadelphia, PA 19103.

This is a Knight Ridder company which contains the full text of news items and feature stories from many newspapers such as the *Detroit Free Press, The Chicago Tribune, The Wall Street Transcript, The Washington Post,* and others.

33. WILSONLINE. The H. W. Wilson Co., 950 University Ave., Bronx, NY 10452-9978.

An online retrieval service offering access to the information in the Wilson Company's printed indexes. Among those of interest are:
Applied Science & Technology Index
Biography Index
Biological & Agricultural Index
Book Review Digest
Business Periodicals Index
Cumulative Book Index—Complete bibliographic records of hardcover and paperback English language books published around the world from January 1982 to date
General Science Index
Readers Guide to Periodical Literature

ENCYCLOPEDIAS, DICTIONARIES, AND HANDBOOKS

To get background or general information and definitions on a particular subject, you may resort to some of the references listed in this section. Encyclopedias include comprehensive treatments, and handbooks or yearbooks can be used for ready reference or concise descriptions of a topic.

34. *ACS Directory of Graduate Research.* Washington, DC: American Chemical Society, 1985. (Biennial)
   Covers institutions offering master's and/or doctoral degrees in chemistry, chemical engineering, and other scientific disciplines.

35. *Aldrich Catalog/Handbook of Fine Chemicals.* 1986/87 ed. Milwaukee, WI: Aldrich Chemical Co., Inc. 1986.
   Lists current prices of organic and inorganic chemicals, biochemicals; computer searches for more current prices are available on request from the company.

36. *Annual Bulletin of Trade in Chemical Products, 1984.* Economic Commission for Europe. New York: United Nations Publications, 1986. 285 p.
   Includes data on import-export of chemical products for twenty-two countries worldwide and covers about sixty-seven commodities.

37. *Annual Review of the Chemical Industry, 1984.* New York: United Nations Publications, 1986. 210 p.
   Includes data on production, import-exports of chemical products, and employment.

38. *Chemcyclopedia 87.* Edited by Joseph H. Kuney. Washington, DC: The American Chemical Society, 1986. 330 p. (Annual)
   The manual of commercially available chemicals. Complete index of all chemicals listed and supplier information.

39. *Chemical Week Buyers Guide, 1986.* New York: McGraw-Hill, Inc. (Annual)
   A complete guide to sources of supply for chemicals and chemical packaging; it contains a chemical company directory, chemical tradenames, and other information.

40. *Chemistry in America, 1876–1976: Historical Indicators.* By A. Thackray et al. Hingham, MA: D. Reidel Publishing Co., dist. by Kluwer Academic Publishers, 1985. 564 p.

41. *Chemsources–U.S.A.* 26th ed. Clemson, SC: Directories Publishing Co., Inc., 1986. (Annual)
   An alphabetical arrangement of chemicals produced or distributed by over 800 firms. The information is obtained from company catalogs, product lists, and computer checklists. There is a separate trade name section and company directory section.

42. *China Chemical Industry 1985/86; World Chemical Industry Yearbook.* Edited by the Scientific and Technical Information Research Institute of the Ministry of Chemical Industry (Beijing, China). New York: VCH Publishers, Inc., 1986. 444 p.
   Gives the status of the Chinese chemical industry and brief descriptions of more than 200 Chinese companies.

43. *College Chemistry Faculties.* 7th ed. Washington, DC: American Chemical Society, 1986. 200 p.
   Information on over 2,000 departments in colleges and universities in the United States and Canada offering instruction in chemistry, biochemistry, chemical engineering, and other related disciplines.

44. *Commercial Atlas & Marketing Guide.* 117th ed. Chicago: Rand McNally & Co., 1987. 589 p.
   Provides statistics and indexes for every populated place in the U.S., railroad listings and maps, zip code maps, and other current economic data used in marketing.

45. *Concise Chemical and Technical Dictionary.* Rev. and enl. 4th ed. Edited by H. Bennett. New York: Chemical Publishing Co., Inc., 1986. 1,271 p.
   Entries from chemistry, biology, physics, medicine, mineralogy, and metallurgy.

46. *Condensed Chemical Dictionary.* 11th ed. By Gessner G. Hawley. Revised by N. Irving Sax and Richard J. Lewis, Sr. New York: Van Nostrand Reinhold Co., 1987. 1,288 p.
   Descriptive information on chemicals and identification of trademarked products used in the chemical industries.

47. *Directory of American Research and Technology.* 20th ed. New York: R. R. Bowker & Co., 1986. 730 p.
   Formerly titled *Industrial Research Laboratories of the U.S.*, this is a compilation of the research and/or development activity at public and private businesses in the U.S. Included are names of key personnel, addresses, and areas of research. Also available on PERGAMON INFOLINE.

48. *Directory of Custom Chemical Manufacturers.* New York: Delphi Marketing Services, Inc., 1985. 161 p.
   This directory presents data on custom chemical manufacturers, cross-indexed by the unit processes and reactions carried out, areas of expertise, and geography.

49. *Directory of World Chemical Producers.* 2d ed. Oceanside, NY: Chemical Information Services, Ltd., 1984. 575 p.
   Coverage consists of 50,000 product listings manufactured by over 5,000 chemical producers in fifty-nine countries around the world. There are two sections: one by product name and the other section by country and producer.

50. *Encyclopedia of Associations.* 22d ed. Edited by Katherine Gruber. Detroit, MI: Gale Research Co., 1987. 4 vols. 3,632 p.
   A guide to national and international organizations in four volumes:
   Vol. 1—National organizations of the U.S.; detailed descriptions of national trade, professional, and other organizations.
   Vol. 2—Geographic and executive indexes covering material in Vol. 1.
   Vol. 3—New associations and projects and updating service.
   Vol. 4—International organizations; detailed descriptions of international trade, professional, and other organizations.

51. *Encyclopedia of Chemical Trademarks and Synonyms.* Edited by H. Bennett. New York: Chemical Publishing Co., Inc., 1981. 3 vols.
   The scope of this work includes chemical compounds and compositions consisting of one or more chemicals and other products produced in the U.S. and in foreign countries.

52. *The Encyclopedia of Chemistry.* 3d ed. Edited by Clifford A. Hampel and Gessner G. Hawley. New York: Van Nostrand Reinhold Co., 1981. 1,182 p.

53. *Europa Yearbook.* 28th ed. London: Europa Publications, Ltd., dist. by Gale Research Co., Detroit, MI, 1987. 2 vols. 3,000 p.
   Basic information on every country with detailed directories and surveys.

54. *Federal Statistical Directory.* 28th ed. By William R. Evinger. Phoenix, AZ: Oryx Press, 1987. 128 p.
   Gives names, titles, addresses and phone numbers of key personnel in statistical programs and other activities of the legislative, judicial, and executive branches of the federal government.

55. *Glossary of Chemical Terms.* 2d ed. Edited by Clifford A. Hampel and Gessner G. Hawley. New York: Van Nostrand Reinhold, Co., 1982. 320 p.
   Chemical terms are explained, and included are industry histories and brief biographies of famous chemists.

56. *Grant and Hackh's Chemical Dictionary.* 5th ed. Edited by Roger Grant and Claire Grant. New York: McGraw-Hill Book Co., 1987. 641 p.
   This new edition presents detailed information on more than 55,000 chemical terms and gives coverage to specialties that were just beginning to be important when the previous edition was published, including biotechnology, immunology science, lasers, and robots.

57. *Handbook of Chemical Synonyms and Trade Names.* 8th ed. Edited by William Gardner. Cleveland, OH: CRC Press, 1978. 769 p.
   A dictionary and commercial handbook containing over 35,000 definitions.

58. *Industrial Chemicals.* 4th ed. Edited by W. L. Faith, Donald B. Keyes and Ronald L. Clark. Revised by F. A. Lowenheim and M. K. Moran. New York: John Wiley & Sons, Inc., 1975. 904 p.

59. *Japan Chemical Directory.* Tokyo, Japan: Japan Chemical Week, The Chemical Daily Co., Ltd., 1985. (Annual)
   Lists manufacturers of chemicals in Japan and includes an appendix with Asian chemical firms.

60. *Japan Company Handbook.* Tokyo, Japan: Oriental Economist, 1985. (Semiannual)
   Covers Japanese corporations listed on the Tokyo, Osaka, and Nagoya Stock Exchanges and lists headquarters location, employee size, sales breakdown, outlook, and other financial information.

61. *JCW Chemicals Guide, 1986/87.* Tokyo, Japan: Japan Chemical Week, The Chemical Daily Co., Ltd., 1986. 558 p.
   A list of chemicals and major Japanese manufacturers and addresses.

62. *Key Chemicals & Polymers.* 7th ed. Washington, DC: American Chemical Society, 1986. 48 p.
   One-page reviews contain data on the economic status and short-term outlook for large volume basic chemical industry products. The articles originally appeared in issues of *Chemical & Engineering News*.

63. *Kirk-Othmer Encyclopedia of Chemical Technology.* 3d ed. New York: John Wiley & Sons, Inc. 26 vols.
   Consists of many articles by recognized experts in the areas of chemical technology, industrial products, natural materials, and processes. Also available online on BRS and DIALOG in a full-text format.

64. *McGraw-Hill Concise Encyclopedia of Science & Technology.* Edited by Sybil P. Parker. New York: McGraw-Hill Book Co., 1984. 2,065 p.
   Extracted from the fifteen-volume *McGraw-Hill Encyclopedia of Science & Technology* (5th ed., 1982), with the addition of several other topics not included in the parent work.

65. *McGraw-Hill Dictionary of Chemical Terms.* Edited by Sybil P. Parker. New York: McGraw-Hill Book Co., 1985. 470 p.

66. *National Trade and Professional Associations of the United States.* 22d ed. Washington, DC: Columbia Books, Inc., 1987. 453 p.
   Includes about 6,250 active national trade and professional associations and labor unions. This directory has an acronym index which is a very useful feature.

67. *OPD Chemical Buyers Directory.* 73d ed. New York: Schnell Publishing Co., Inc., 1986. (Annual)
   Lists sources of supply for chemicals and related process materials and appears once a year as part of a subscription to the *Chemical Marketing Reporter*.

68. *Opportunities in Chemical Distribution; Dynamics of a Growing Industry.* Fairfield, NJ: Charles H. Kline & Co., 1985. 454 p.
   This source provides distributor profiles of the leading U.S. chemical distributors and includes analyses of the industry and end use industries.

69. *Profiles of U.S. Chemical Distributors.* 1st ed. Fairfield, NJ: Charles H. Kline & Co., Inc., 1981. 238 p.
    Comprehensive profiles on over 900 U.S. chemical distributors including headquarters and branches. It also gives overall industry statistics such as sales, geographic concentration, and trends.

70. *Research Centers Directory 1987.* 11th ed. Detroit, MI: Gale Research Co., 1986. 1,700 p. (Annual)
    University-related and nonprofit research organizations involved in research on a permanent and continual basis in all areas of study.

71. *Riegel's Handbook of Industrial Chemistry.* 8th ed. Edited by James A. Kent. New York: Chemical Publishing Co., Inc., 1983. 1,008 p.
    Contains twenty-five chapters on various topics relating to the manufacture of chemicals and the design and operation of chemical plants.

72. *Specialty Chemicals Handbook.* Edited by Japan Chemical Week. Tokyo, Japan: The Chemical Daily Co., Ltd., 1984. 337 p.
    Gives import-export statistics, production, market price, and Japanese manufacturers for organic intermediates.

73. *Statistical Yearbook, 1983/84.* 34th issue. New York: United Nations Publications, 1986. 900 p.
    Data for over 250 countries on economic and social subjects including population, manufacturing, etc.

74. *Thesaurus of Chemical Products.* Edited by Michael Ash and Irene Ash. New York: Chemical Publishing Co. 2 vols. Vol. I: Generic-to-Tradename, 1985. 400 p.; Vol. II: Tradename-to-Generic, 1986. 400 p.
    Formerly called *Source Handbook of Generic Products.*

INDEXES, ABSTRACTS, AND CONTINUING SERVICES

These references may be used to retrieve information about new developments on specific topics. Many have abstracts that are very descriptive. Increasingly, there is an online access to many of these sources, and this is indicated.

Indexes and Abstracts

75. *Business Periodicals Index.* Bronx, NY: H. W. Wilson Co., 1958–. (Monthly)
    A single alphabet subject index of over 300 business magazines with many cross references and a separate index to book reviews.

76. *Chemical Industry Notes.* Columbus, OH: American Chemical Society, Chemical Abstracts Service, 1971–. (Weekly)
    Available also on DIALOG.

77. *CPI Digest.* Louisville, KY: CPI Information Services, Inc., Division of K&M Publications, Inc., 1974–. (Monthly)
Surveys new marketing and technical developments in the chemical process industries. Covers coatings, adhesives, inks, plastics, rubber, fibers, pigments, and polymers.

78. *Polymers/Ceramics/Composites Alert.* Metals Park, OH: Materials Information. (Monthly)
This abstracting publication provides international coverage of business developments for materials industries.

79. *PROMT.* Cleveland, OH: Predicasts, Inc. (Monthly, Quarterly, Annually)
PROMT is an acronym for *Predicasts Overview of Markets and Technologies.* Articles are abstracted from the worldwide literature for several subject areas including agriculture, food, packaging, paper, plastics, drugs and toiletries, chemicals, electronics, and energy. Also available on DIALOG.

80. *PTS Funk & Scott Indexes.* Cleveland, OH: Predicasts U.S., Europe, and International Indexes, (Weekly, Quarterly, Annually)
Cumulative indexes to published articles and reports by Standard Industrial Classification number, by company name, and by country. Also available on DIALOG.

81. *Recent Additions to Baker Library.* Boston: Harvard Business School, Baker Library. (Monthly)

82. *Technology Update.* Cleveland, OH: Predicasts, Inc. (Weekly)
Reviews developments in technology for major industry categories taken from journals, government reports, research studies, and other documents.

83. *What's New in Advertising and Marketing.* New York: Special Libraries Association. (10/yr.)
Lists current materials in advertising, marketing, and communications.

Continuing Services

Continuing services are usually provided on a contractual basis and include the output of various research firms who gather data and publish it in printed or electronic formats. Services from these companies sometimes include consulation as a part of the contractual agreement. The subjects covered can be very specific, such as those covering specific products or industries, or very broad, such as management, corporate strategy, etc. The following listing includes names, addresses, and major publications or services.

84. Arthur D. Little Decision Resources. Arthur D. Little, Inc., 17 Acorn Park, Cambridge, MA 02140.

A research, development, and management consulting firm that provides a group of interpretive information services, including consulting, meetings, publications and database access. The principal services include the *Health Care Industry Service, The Forefront Biotechnology Service, The World Telecommunications Service, Integrated Information Systems Service,* and *InfoTran.*

85. Battelle. 505 King Ave., Columbus, OH 43201-2693.

This large international firm specializes in materials and technology research. It provides in-depth patent and literature searches and extensive testing in its own laboratories. *The B-Tip Program: Battelle Technical Inputs to Planning* provides several resources—briefings and consultations, meetings, and publications. The publications consist of reports and reviews and a newsletter titled *Technology Sensor,* which is published six times a year to profile emerging technologies.

86. *CAPPS–Chemicals and Polymer Production Statistics.* New York: McGraw-Hill, DATA RESOURCES/Chemical Week.

A monthly analysis of U.S. chemical and polymers production and trade data. Analyzes and reports production and foreign trade data on forty-seven chemical raw materials and polymers.

87. Chem Systems, Inc. 303 S. Broadway, Tarrytown, NY 10591.

A marketing research firm which provides in-depth reports, seminars and consulting services, techno-economic analysis of chemical and polymer industries, industrial and commodity chemicals. One of their major services to the chemical industry is titled, *Process Evaluation and Research Planning Service (PERP).* It consists of major or in-depth reports and topical reports chosen from a variety of fields, including polymers, specialty chemicals, fine chemicals, commodities, advanced materials, and biotechnology.

88. *Chemical Profiles.* New York: Schnell Publishing Co.

Brief summaries of 150 major commodity chemicals; updated approximately every three years. These are issued every quarter in packages of thirteen.

89. The Conference Board. 845 Third Ave., New York, NY 10022.

Provides a management information service and does research in the fields of economic conditions, marketing, finance, personnel administration, and international activities. Publishes research reports, periodicals, and provides information on request.

90. Dun's Marketing Services. Dun & Bradstreet Corp., 3 Century Dr., Parsippany, NJ 07054.

Dun's Marketing Services provides the following three sources that are helpful to the chemical industry: *The Million Dollar Directory Series,* 5 vols.—Lists companies with a net worth of $500,000+ by alpha, geographically, and by industry or SIC code. Also available online with DIALOG. *Principle International Business*—Information on leading companies in 133 countries throughout the free world. Also available online with DIALOG. *Who Owns Whom*—Four geographic editions trace the structure and ownership of multinational corporations.

91. *Mannsville Chemical Product Synopses.* Mannsville Chemical Products Corp., Box 232, Cortland, NY 13045.
Current two-page chemical profiles on some basic commodities. Covers about 210 and updates at least half of these each year. This is a reporting service which provides information on a particular chemical emphasizing market, price, and availability aspects as well as technological and environmental aspects.

92. *Moody's Manuals and News Reports.* New York.
This reference service provides profiles of publicly held companies in U.S. industries: banking and finance, transportation, utilities, municipalities and governments, and international corporations and financial institutions. Also available online with DIALOG.

93. *The Rome Report.* Leading National Advertisers, subsidiary of Interactive Market Systems, 136 Madison Ave., 5th Fl., New York, NY 10016. (Semiannual)
This service provides page and dollar expenditures for all advertisers in over 580 trade and business publications from 90 classifications.

94. SRI, International. 333 Ravenswood Ave., Chemical Industries Center, Menlo Park, CA 94025.
Performs techno-economic research on a multiclient and private basis for commercial and industrial clients. Provides the following continuing services: *Business Intelligence Program*—Scans, monitors and researches economic, technological and social trends likely to impact on corporate strategy. Also available to subscribers on DIALOG. *Chemical Economics Handbook*, 35 vols.—Provides detailed marketing information on some chemical commodities such as production, producers, end uses, foreign trade, prices, and summaries. The full text of this output is also available online with SDC. *Directory of Chemical Producers, U.S. & Western Europe*—Published annually; gives company names, addresses, chemicals produced, production sites, and comments. *Process Economics Program*, 180 vols.—Provides plant costs and costs of making a chemical commodity by a particular method or process. *Specialty Chemicals Update Program*, 10 vols.—Provides detailed analyses of selected segments of the world's specialty chemical business. Each report gives information on end use markets, a description of a particular specialty chemical business and an analysis of the worldwide markets with five-year projections. Each report is updated approximately once every three years. *World Petrochemicals*—Covers 200 petrochemicals, including basic hydrocarbons, intermediates, plastics, fibers, elastomers, and solvents. The data is presented by country and region. The service also includes a price update (monthly). Data is accessible online through WP DATA. Note: SRI's Chemical Industries Center also offers inquiry privileges with many of its services to provide additional and current information on chemical producers and products.

95. Standard & Poor's Corp. McGraw-Hill, Inc., 25 Broadway, New York, NY 10004.

This firm provides business and financial information on companies and industries in a variety of formats. Some of the general services are: *Industry Surveys*—Business and economic information on all major U.S. industries (twenty-two individual surveys). Basic surveys are published annually, and other data is published periodically. *Register of Corporations, Directors and Executives*, Annual—This publication consists of three volumes: Vol I, *Corporate Listings*; Vol II, *Biographical Profiles of Executives*; and Vol III, *Geographical and SIC Code Indexes*. *Statistical Service*, Monthly—Over 1,000 economic and financial series.

NEWSPAPERS, JOURNALS, AND NEWSLETTERS

Newspapers, journals, and newsletters perform a current awareness function by providing information on news and trends in industries. Most of the format of the data is short and timely, and many of the publications regularly provide special issues that give forecasts or reviews of the industry.

96. *Advertising Age.* Chicago: Crain Communications, Inc. (Weekly)

Publishes several surveys of interest. One is "100 Leading National Advertisers," which has profiles and facts and figures on marketing operations of each company.

97. *Business Marketing.* Chicago: Crain Communications, Inc. (Monthly)

This source features information that can be used in advertising and selling to business and industry.

98. *Business Week.* New York: McGraw-Hill, Inc. (Weekly)

This popular magazine focuses on topics that cover management, technology, companies, the economy, and international business. Also published as regular features are lists of leading companies arranged by industry on such topics as R & D and Corporate Performance.

99. *Chemical & Engineering News.* Washington, DC: American Chemical Society. (Weekly)

Special issues include:
January—R&D Spending
February—Chemical Industry Quarterly Earnings
March—Productivity
April—Top 50 Chemical Products
May—ACS Annual Report
   Top 100 Chemical Producers
   30 Top U.S. Chemical Companies
   Chemistry Graduates
June—Facts and Figures for the Chemical Industry
October—Annual Employment Outlook for Scientists and Engineers
December—World Chemical Outlook
   Chemical Capital Spending

100. *Chemical Business.* New York: Schnell Publishing Co. (Monthly)
Published every fourth week along with the regular issue of the *Chemical Marketing Reporter* as an editorial supplement.

101. *Chemical Economy and Engineering Review (CEER).* Tokyo, Japan: Chemical Economy Research Institute. (Monthly)

102. *Chemical Engineering.* New York: McGraw-Hill Publications, Inc. (Fortnightly)

103. *Chemical Industries Newsletter.* Menlo Park, CA: SRI, International.
Published six times a year. Contents are drawn from current research of the chemical centers and other SRI programs of interest.

104. *Chemical Industry Update: North America.* Cleveland, OH: Predicasts. (Weekly)
A weekly review of trends in North American chemical processing industries for sales and marketing executives.

105. *Chemical Insight.* London: Hyde Chemical Publications, Ltd. (Semimonthly)
Perspectives on the international chemical industry. Fall issues have sales rankings and R & D expenditures for individual companies.

106. *Chemical Marketing & Management.* Staten Island, NY: Chemical Marketing Research Association. (Quarterly)
Published quarterly by the Chemical Marketing Research Association. The coverage purports to be broad in scope and will focus on industry activity and strategy, specific product/market areas, and other items of interest.

107. *Chemical Marketing Reporter.* New York: Schnell Publishing Co. (Weekly)
Gives current prices of chemicals and chemical profiles. Special features of regular issues include Chemicals Outlook, Detergents, Petrochemicals, Specialties, Beauty Chemicals, Coatings, and Chemicals Shipping.

108. *Chemical Spotlight.* Ridgewood, NJ: Whitmore Associates. (Weekly)
A current awareness service for executives in the chemical industry. Published every Friday.

109. *Chemical Times & Trends.* Washington, DC: Chemical Specialties Manufacturers Association. (Quarterly)

110. *Chemical Week.* New York: McGraw-Hill, Inc. (Weekly)
Some issues of interest are:
Adhesives—A special advertising section
Chemical Week 300—Top 300 companies by sale
Coatings—A special advertising section Forecast
Plastics—A special advertising section

State Rankings in Chemical Production (9/25/85)

111. *Chemistry and Industry.* London: Society of Chemical Industry. (Monthly)

112. *Chemscope.* Sutton, Surrey, England: European Chemical News, Enterprise Publishing; Business Press, International.
Presented with *European Chemical News* as a supplement.

113. *CHEMTECH.* Washington, DC: The American Chemical Society. (Monthly)
Subtitled the "Innovator's Magazine," it purports to bridge the gap between theory and practice by featuring topics vital to the innovative process.

114. *Chemweek Newswire.* New York: McGraw-Hill, Inc. (Daily)
A service of *Chemical Week Magazine* available in a wire edition and a mail edition. The mail edition is also available on NEXIS. Contents include items of interest to industries involved in the chemical processes and includes important events that have occurred in the industry that day.

115. *CPI Purchasing.* Boston: Cahners Publishing Co. (Monthly)
Includes a business survey, a section called "Chemforecast" (on individual commodities), and a section on chemical prices. Supplier profiles with detail on products, plant sites, and capacities are a regular feature.

116. *Dun's Business Month.* New York: Technical Publishing Co. (Monthly)

117. *European Chemical News.* Sutton, Surrey, England: Enterprise Publishing; Business Press International, Ltd. (Weekly)
News summaries and market trend information; includes a bulk chemical price report.

118. *Forbes.* New York: Forbes, Inc. (Biweekly)
Gives timely information on industries and companies. Special issues include:
January—Annual Report on American Industry.
May—Directory Issue.
July—Special Report on International Business
August—Mutual Funds Survey
September—The Forbes 400

119. *Fortune.* New York: Time, Inc. (Biweekly)
Articles feature stories on business and management subjects. Special issues are devoted to rankings of corporations, the most famous being "The Fortune 500," which lists information on U.S. corporations ranked by sales.

120. *Handbook of Basic Economic Statistics.* Washington, DC: Economic Statistics Bureau. (Monthly)
  A manual of basic economic data on industry, commerce, labor, and agriculture in the U.S.

121. *Industrial Chemical News.* New York: Bill Communications, Inc. (Monthly)

122. *Industry Week.* Cleveland, OH: Penton Publishing Co., Inc. (Biweekly)

123. *International Financial Statistics.* Washington, DC: International Monetary Fund. (Monthly)

124. *Japan Chemical Week.* Tokyo, Japan: The Chemical Daily Co., Ltd. (Weekly)

125. *Journal of Commerce and Commercial.* New York: Twin Coast Newspapers, Inc. (5/week)
  Includes a section on chemicals and plastics and a spot chemicals prices table with spot quotations of list prices of suppliers. Other regular features are petroleum prices and a plastics news section.

126. *Main Economic Indicators.* Washington, DC: OECD Publications and Information Center. (Monthly)
  Statistics of member countries including GNP, industrial production, construction, wholesale and retail sales, employment, wages, prices, finance, foreign trade, and balance of payments.

127. *Manufacturing Chemist* (formerly *Chemical Age*). London: Morgan-Grampian, Ltd. (Monthly)
  Covers production, packaging, and marketing of general chemicals, petrochemicals, pharmaceuticals, household chemicals, cosmetics and toiletries, and aerosol products.

128. *Marketing News.* Chicago: American Marketing Association. (Biweekly)

129. *Marketing Times.* Cleveland, OH: Sales and Marketing Executives, International. (Bimonthly)

130. *Monthly Bulletin of Statistics.* New York: United Nations Publications.
  Monthly series of world statistics on seventy-four subjects from over 200 countries.

131. *New Scientist.* Elmont, NY: IPC Magazines, Ltd.; Publications Expediting, Inc. (Weekly)

132. *OECD Observer.* Washington, DC: OECD Publications and Information Center. (Bimonthly)
Information on the activities of OECD; features articles on economic affairs, energy, social affairs, the environment, multinational enterprises, science and technology, and financial markets.

133. *Performance Chemicals.* European Chemical News, Vol. 1, No. 1, May, 1986. Sutton, Surrey, England: Business Press International, Ltd. (Quarterly)
A new quarterly developed from ECN's *Specialty Chemscope Supplements.* Covers key developments in the fine and specialty chemicals area.

134. *Purchasing World.* Solon, OH: International Thomson Industrial Press, Inc. (Monthly)
Designed to help manage the business of buying, this publicaiton contains news of interest to purchasing personnel. It also includes business and economic news and current prices of chemicals and other materials.

135. *The Report on Performance Materials.* Washington, DC: McGraw-Hill, Inc. (26/year)
Covers composites, advanced ceramics, resins, fibers, and specialty metals.

136. *S and MM Sales & Marketing Management.* New York: Bill Communications, Inc. (16/year)
Special issues feature these topics:
February—Survey of Selling Costs
April—Survey of U.S. Industrial & Commercial Buying Power; Totals for four-digit SIC industries and state and county SIC totals.
July—Survey of Buying Power: Part I—Statistics on population, households, income, and retail sales for U.S. metro areas, counties, and cities.
October—Survey of Buying Power: Part II—Five-year projections; TV and newspaper markets.

137. *Scientific Meetings.* San Diego, CA: Scientific Meetings Publications. (Quarterly)
Covers national, international, and regional meetings, symposia, colloquia, and institutes offered by scientific, technical, medical, health, engineering, and management organizations worldwide.

138. *Wall Street Journal.* New York: Dow Jones & Co. (Daily)

139. *Wall Street Transcript.* New York: Wall Street Transcript Corp.. (Weekly)
The full text of selected investment analyst/brokerage house reports on companies and industries. One of the regular features is the round table discussions of industries and companies involved in those industries. Also available online in full text on VU-TEXT.

U.S. GOVERNMENT PUBLICATIONS

Much government data is authoritative and is published as reports, statistical publications, periodicals and news releases by agencies, bureaus, or departments. Sought-after facts such as consumption, production, population, end uses and other statistics can be found in many of these sources. Some of the data is also available online or on tapes and in microforms.

140. U.S. Department of Commerce. *Business America.* Washington, DC: Government Printing Office. (Fortnightly)

141. U.S. Department of Commerce. Bureau of Economic Analysis. *Survey of Current Business.* Washington, DC: Government Printing Office. (Monthly)

142. U.S. Department of Commerce. Bureau of the Census. *Census of Manufactures, 1982.*
    This economic census is taken every five years and gives data on manufacturing establishments in the form of reports on industries, geographic areas, and major subjects.

143. U.S. Department of Commerce. Bureau of the Census. *County and City Data Book; A Statistical Abstract Supplement.* 10th issue. Washington, DC: Government Printing Office, 1983. 996 p.
    Detailed information on regions, divisions, states, counties, metropolitan areas, cities; those areas not generally found in the *Statistical Abstract of the U.S.*

144. U.S. Department of Commerce. Bureau of the Census. *Current Industrial Reports.* Washington, DC: Government Printing Office.
    These reports give current data on production, inventories, and orders for 5,000 products. The frequency of the publications include monthly, quarterly, annual, and biennial reports. The reports are based on categories in the Standard Industrial Classification Codes: Apparel and Leather; Chemicals; Rubber and Plastic; Intermediate Metal Products; Lumber; Furniture and Paper Products; Machinery and Equipment; Primary Metals; Processed Foods; Stone, Clay, and Glass Products; and Textile Mill Products.

145. U.S. Department of Commerce. Bureau of the Census. *Historical Statistics of the United States Colonial Times to 1970.* Washington, DC: Government Printing Office, 1975. 2 vols.
    Retrospective compilation of data. Categories are similar to those in the *Statistical Abstract of the U.S.* Indexed by subject and time period.

146. U.S. Department of Commerce. Bureau of the Census. *State and Metropolitan Area Data Book; A Statistical Abstract Supplement.* 2d issue. Washington, DC: Government Printing Office, 1982.
    Geographic detail on metropolitan areas (SMSAs), divisions, states. Source notes and explanations provide bibliographic references.

147. U.S. Department of Commerce. Bureau of the Census. *Statistical Abstract of the United States, 1987.* 107th ed. Washington, DC: Government Printing Office, 1986. 960 p. (Annual)
   A compilation of statistics from U.S. agencies and some professional and trade associations. Well indexed and contains data on more than thirty topics.

148. U.S. Department of Commerce. Industry and Trade Administration. Office of Field Operations. *Commerce Business Daily.* Washington, DC: Government Printing Office.
   Synopsis of U. S. government proposed procurement.

149. U.S. Department of Commerce. International Trade Commission. *U.S. Industrial Outlook, 1987.* Washington, DC: Government Printing Office, 1987. 660 p. (Annual)
   Gives a detailed account of the industrial economy of the United States and analyzes over 350 industries. The data is organized by the Standard Industrial Classification Code.

150. U.S. Executive Office of the President. Council of Economic Advisers. *Economic Indicators.* Washington, DC: Government Printing Office. (Monthly)
   Data on prices, wages, production, business activity, purchasing power, credit, money, and federal finance.

151. U.S. Federal Reserve System. Board of Governors. *Federal Reserve Bulletin.* Washington, DC: Government Printing Office. (Monthly)

152. U.S. International Tariff Commission. *Tariff Schedules of the United States Annotated, 1987.* USITC Publication 1910. Washington, DC: Government Printing Office, 1986. (Annual)
   A subscription service used in classifying imported merchandise for rate of duty and statistical purposes.

153. U.S. International Trade Commission. *Chemical Industry Growth in Developing Countries and Changing U.S. Trade Patterns.* USITC Publication 1780. Report on Investigation No. 332-198 under Section 332(b) of the Tariff Act of 1930. Washington, DC: U.S. International Trade Commission, 1985. 250 p.
   Data on the position of the United States and the developing nations in the world market for chemicals are presented. Demand, production, consumption, and trade and their possible effects on U.S. chemical trade are analyzed.

154. U.S. International Trade Commission. *The Shift from U.S. Production of Commodity Petrochemicals to Value Added Specialty Chemical Products and the Possible Impact on U.S. Trade.* USITC Publication 1677. Report on Investigation No. 332-183 under Section 332(b) of the Tariff Act of 1930. Washington, DC: U.S. International Trade Commission, 1985.

This report presents information and data from diverse sources on the changing competitive position of the United States and other nations in the world petrochemical market.

155. U.S. International Trade Commission. *U.S. Production and Sales of Synthetic Organic Chemicals, 1985.* USITC Publication 1745. Washington, DC: Government Printing Office, 1986. (Annual)
Reports the domestic production and sales of synthetic organic chemicals and the raw materials from which they are made. The report consists of fifteen sections, each covering a specified group of organic chemicals. The data are supplied by over 750 producers whose names and addresses are included in a directory within the report.

156. U.S. Office of the Federal Register. *The Code of Federal Regulations.* Washington, DC: Government Printing Office. (Annual)
This is divided into fifty titles and is an annual codification of the general and permanent rules published in *The Federal Register.* The code is updated by the individual issues of *The Federal Register.*

157. U.S. Office of the Federal Register. *The Federal Register.* Washington, DC: Government Printing Office. (Daily)
Issued each federal working day, this provides presidential documents, proposed rules and regulations, and documents required to be published by statute.

158. U.S. Office of the Federal Register. *The United States Government Manual, 1986/87.* Washington, DC: Government Printing Office, 1986. (Once every two years)
The official handbook of the federal government, this publication provides comprehensive information on governmental agencies and gives their purpose, history, program, and activities. The manual is published as a special edition of *The Federal Register.*

Categorical Arrangement

AGRICULTURAL CHEMICALS, FOOD AND BEVERAGES

Because chemicals play a complex role in the production and processing of food, the sources listed in this section will cover several industries connected with these important commodities. Agricultural chemicals consist of fertilizers and pesticides such as herbicides, insecticides, and fungicides. The manufacturing of these products is considered essential for successful growth. The food and beverage industry is composed of several large groups, such as meat and dairy products, preserved fruits and vegetables, grain mill products, bakery products, fats and oils, and beverages. Most of the chemicals used in food processing are food additives, which are used to perform different functions, such as flavoring, thickening, stabilizing or emulsifying, sweetening, coloring, preserving or canning.

Biographical Directories

159. *American Journal of Agricultural Economics Handbook Directory.* Ames, IA: American Agricultural Economics Association, 1984. (Every four years)

Electronic Retrieval Systems

160. AGRIBUSINESS U.S.A. Des Moines, IA: Pioneer Hi-Bred International, Inc.
   Indexes and abstracts from more than 300 business and government publications. Also includes the full text of USDA statistical reports published since January 1986. Also available on DIALOG and updated every two weeks.

161. AGRICOLA. Washington, DC: National Agricultural Library.
   An acronym for Agricultural On-Line Access, this database is produced by the National Agricultural Library. Citations to the published literature in every field of agriculture will be found in the system. Representative topics are agricultural marketing, fertilizers, food and feeds, pesticides, and plant genetics. Also available on DIALOG.

162. AGRIDATA NETWORK. Milwaukee, WI: Agridata Resources, Inc.
Data in this information bank include reports on the weather; agricultural news and markets; and government, university, and association activities.

163. CRIS. Washington, DC: Current Research Information System.
Covers ongoing and recently completed agricultural research projects sponsored by state and federal organizations. Also available on DIALOG.

164. CURRENT BIOTECHNOLOGY ABSTRACTS. London: The Royal Society of Chemistry. (Monthly)
Covers scientific, technical, and commercial literature in biotechnology, pharmaceuticals, food, and agriculture. Also available on PERGAMON INFOLINE.

Encyclopedias, Dictionaries, Handbooks

165. *Agri Marketing; Marketing Services Guide Issue.* Skokie, IL: Century Communications, Inc., 1986. (Annual)
Lists agricultural companies, agricultural advertisers, marketing services firms, broadcasting stations, and agricultural associations.

166. *The Almanac of the Canning, Freezing, Preserving Industries.* 71st ed. Westminster, MD: Edward E. Judge & Sons, Inc., 1986. 690 p. (Annual)
A source of basic references for these industries.

167. *Beverage Industry Annual Manual, 1987.* Cleveland, OH: Magazines for Industry, Inc. (Annual)
Provides statistics on the industry. Covers beverage plants and has a buyer's guide and brand name directory.

168. *Beverage World Data Bank, 1986/87.* Dayton, OH: Keller International Publishing Corp., 1986. 396 p. (Annual)
Provides suppliers, equipment, conventions, services, rankings of beverage companies by sales and brands, consultants, associations, and other information.

169. *Commodity Year Book, 1987.* Jersey City, NJ: Commodity Research Bureau, Knight Ridder Business Information, Inc. 296 p. (Annual)
Provides supply and demand data for 110 basic commodities and future markets. Includes many tables and price charts with some as far back as 1920.

170. *The Directory of the Canning, Freezing, Preserving Industries.* Westminster, MD: James J. Judge, Inc., 1986–87. 600 p.
Data on companies in this industry is presented in five basic sections: alphabetical by country, geographical, by product, brand, and an association listing.

171. *Farm Chemicals Handbook, 1987.* Willoughby, OH: Meister Publishing Co., 1986. (Annual)
   Manufacturers and suppliers of chemicals used in agribusiness.

172. *Feedstuffs—Reference Issue, 1986.* Minneapolis, MN: Miller Publishing Co. (Annual)
   Listings of suppliers and equipment for the feed and grain industries.

173. *Food Ingredients Directory.* Hastings-on-Hudson, NY: 1971–. (Quarterly updates)
   A guide to manufacturers and suppliers of ingredients and chemicals used by the food processing industry.

174. *Food Processing Ingredients, Equipment & Supplies Guide & Directory.* 1988 ed. Chicago: Putman Publishing Co., 1987. 500 p.
   Lists food ingredients and manufacturers. This source also includes a guide to national associations, a guide to government agencies, plant sites and food laboratories.

175. *Quick Frozen Foods' Directory of Frozen Food Processors and Buyers Guide, 1987.* 39th ed. New York: Frozen Food Digest, Inc., 1987. 350 p. (Annual)

Indexes, Abstracts, and Continuing Services

176. *Agriculture and Food. Abstract Newsletter.* Springfield, VA: U.S. National Technical Information Service. (Weekly)

177. *AGRINDEX.* Rome, Italy: Food and Agriculture Organization of the United Nations. Distributed in the U.S. by UNIPUB, New York. (Monthly) 1975–.
   Provides international information on agricultural economics and technology.

178. *Commodity Yearbook Statistical Abstract Service.* Jersey City, NJ: Commodity Research Bureau, Inc. (3/year)

179. *Food Science and Technology Abstracts.* New York: UNIPUB. Also available on DIALOG. (Monthly) 1969–.

180. *Foods Adlibra.* Minneapolis, MN: Foods Adlibra Publications, General Mills, Inc. Also available on DIALOG. (Monthly) 1974–.

Newspapers, Journals, and Newsletters

181. *Agmarketer.* Yakima, WA: Columbia Publishing Co. (Monthly)
   January—Chemical Fertilizer Issue.

182. *Agrichemical Age.* San Francisco, CA: California Farmer. (10/year)
   Provides uses and applications of agricultural chemicals.

183. *Agri Marketing.* Skokie, IL: Century Communications. (Monthly)

184. *American Journal of Agricultural Economics.* Ames, IA: American Agricultural Economics Association. (5/year)

185. *Beverage Industry.* Cleveland, OH: Magazines for Industry, Inc. (Monthly)

186. *Beverage World.* Great Neck, NY: Keller International. (Monthly)
Special issues include:
June—Annual Packaging Issue.
July—Top 100 Beverage Companies.

187. *Biotechnology News.* Maplewood, NJ: CTB International Publishing Co. (Biweekly)

188. *Farm Chemicals.* Willoughby, OH: Meister Publishing Co. (Monthly)

189. *Farm Journal.* Philadelphia, PA: Farm Journal, Inc. (Monthly)

190. *Farm Supplier.* Washington, DC: Food Chemical News, Inc. (Monthly)
Special issues are:
January—Farm Chemicals/Herbicides
March—Farm Chemicals/Insecticides
September—Animal Healthcare
October—Hardware, Lawn and Garden
December—Directory Issue

191. *Feedstuffs.* Minnetonka, MN: Miller Publishing Co. (Weekly)

192. *Food & Beverage Marketing.* New York: U.S. Business Press, Inc. (Monthly)

193. *Food Chemical News.* Washington, DC: Food Chemical News, Inc. (Weekly)

194. *Food Development.* New York: Magazines for Industry, Inc. (Monthly)

195. *Food Engineering.* Radnor, PA: Chilton Co. (Monthly)
Special issues include:
May—Ingredients Report
July—Directory of U.S. Food Plants

196. *Foodlines.* Washington, DC: Food Research & Action Center. (Monthly)

197. *Food Processing.* Chicago: Putman Publishing Co. (Monthly)
Special issues include:
January/April/July/October—Foods of Tomorrow—New Products & Ingredients
February—Food Industry Outlook
December—Top 100 Food Companies

198. *Food Technology.* Chicago: Institute of Food Technologists. (Monthly)

199. *Frozen Food Digest.* New York: Frozen Food Digest, Inc. (Quarterly)

200. *Green Markets.* New York: McGraw-Hill, Inc. (Weekly)
Provides an analysis of major chemicals used in fertilizers; also gives market prices.

201. *Kiplinger Agricultural Letter.* Washington, DC: Washington Editors, Inc. (Fortnightly)

202. *Lawn and Garden Marketing.* Overland Park, KS: Intertec Publishing Corp. (10/year)

203. *Lawn Care Industry.* Middleburg Heights, OH: Harcourt Brace Jovanovich, Inc. (Monthly)
The June issue presents the state of the industry.

204. *Modern Brewery Age.* Norwalk, CT: Business Journals, Inc. (Weekly)
Special issues include:
February/March—Annual Statistical and Yearend Roundup
Spring—Blue Book; Annual Directory

205. *Pest Control.* Cleveland, OH: Harcourt Brace Jovanovich, Inc. (Monthly)

206. *Prepared Foods.* Chicago: Gorman Publishing Co. (13/year)

207. *Quick Frozen Foods.* New York: Harcourt Brace Jovanovich, Inc. (Monthly)
October or November—Frozen Food Almanac Statistics

208. *Snack Food.* New York: Harcourt Brace Jovanovich, Inc. (Monthly)
Special issues include:
May/November—Packaging
October—Flavors and Ingredients

209. *Weed, Trees and Turf.* Cleveland, OH: Harcourt Brace Jovanovich, Inc. (Monthly)
Special issues include:
March—Weed Control Guide
May—Insect Control Guide
November—Fertilizer Guide

210. *World Food & Drink Report.* Washington, DC: King Communications Group, Inc. (Biweekly)

211. *Yard & Garden.* Ft. Atkinson, WI: Johnson Hill Press, Inc. (10/year)

U.S. Government Publications

212. U.S. Department of Agriculture. Crop Reporting Board. *Crop Production.* Washington, DC: Government Printing Office. (Monthly, Seasonal, and Annual)

213. U.S. Department of Agriculture. Crop Reporting Board. Statistical Reporting Service. *Agricultural Prices.* Washington, DC: Government Printing Office. (Monthly)

214. U.S. Department of Agriculture. Economic Research Service. *Agricultural Exports Outlook and Situation.* Washington, DC: Government Printing Office. (Quarterly)

215. U.S. Department of Agriculture. Economic Research Service. *World Agricultural Supply and Demand Estimates.* Washington, DC: Government Printing Office. (Monthly)

216. U.S. Department of Agriculture. Economic Research Service. *World Agriculture: Outlook and Situation.* Washington, DC: Government Printing Office. (Quarterly)

217. U.S. Department of Agriculture. Economics Management Staff. *National Food Review.* Washington, DC: Government Printing Office. (Quarterly)

218. U.S. Department of Agriculture. Economics Management Staff. Information Division. *Agricultural Economics Research.* Washington, DC: Government Printing Office. (Quarterly)

219. U.S. Department of Agriculture. Economics Management Staff. Information Division. *Agricultural Outlook.* Washington, DC: Government Printing Office. (11/year)

220. U.S. Department of Agriculture. Economics Management Staff. Information Division. *Foreign Agricultural Trade of the United States.* Washington, DC: Government Printing Office. (Bimonthly)

30 *Chemical Industries: An Information Sourcebook*

221. U.S. Department of Agriculture. Foreign Agricultural Service. Information Division. *Farmline.* Washington, DC: Government Printing Office. (Monthly)

222. U.S. Department of Agriculture. Foreign Agricultural Service. Information Division. *Foreign Agriculture.* Washington, DC: Government Printing Office. (Monthly)

223. U.S. Department of Agriculture. Statistical Reporting Service. *Agricultural Statistics.* Washington, DC: Government Printing Office. (Annual)

224. U.S. Department of Commerce. Bureau of the Census. *Census of Agriculture, 1982.* Washington, DC: Government Printing Office.
   Reports are based on data collected every five years and include statistics on farms, livestock, poultry, crops, irrigation, farm products, and other data.

CLEANING PREPARATIONS, COSMETICS AND TOILETRIES

The major product categories covered in this section include soaps and detergents, polishes and sanitation goods, surface-active agents, and cosmetics and toiletries. Some of the basic business sources of information on the personal grooming industry and the household and industrial cleaning products industry will be featured. Included in cosmetics and toiletries are deodorants and antiperspirants, fragrances, hair care, makeup, nail care, oral care and skin care. Cleaning products include household bleaches, specialty cleaners and disinfectants, and polishing preparations.

Encyclopedias, Dictionaries, and Handbooks

225. *McCutcheon's Emulsifiers and Detergents, North American Edition.* Glen Rock, NJ: McCutcheon's Publications, 1987. 326 p.
   Surfactant materials described by trade name, identity, manufacturer, type, and application.

226. *Soap/Cosmetics/Chemical Specialties Blue Book, 1986.* New York: MacNair-Dorland Co. 1987. 140 p. (Annual)
   Sources of raw materials, equipment, and services for the soap, cosmetic, and chemical specialties industry.

Newspapers, Journals, and Newsletters

227. *American Laundry Digest.* Chicago: American Trade Magazines, Inc. (Monthly)

228. *Cleaning Management.* Irvine, CA: Harris Communications. (Monthly)
   In January a Buyers Guide Issue is featured.

229. *Coinamatic Age.* New York: Coinamatic Age. (Bimonthly)

230. *Cosmetic World News.* London: World News Publications. (6/year)

231. *Cosmetics & Toiletries.* Wheaton, IL: Allured Publishing Corp. (Monthly)

232. *Drug and Cosmetic Industry.* New York: Harcourt Brace Jovanovich, Inc. (Monthly)
   The August or October issue features a Drug and Cosmetic Catalog.

233. *Household & Personal Products Industry.* Ramsey, NJ: Rodman Publishing Corp. (Monthly)

234. *The Rose Sheet.* Chevy Chase, MD: F-D-C Reports, Inc. (Weekly)
   Reports feature news on the toiletries, fragrances, skin care, and related industries.

235. *Product Marketing and Cosmetic and Fragrance Retailing.* New York: Business Press, Inc. (Monthly)
   The November issue features Fragrance Forecasts.

236. *Sanitary Maintenance.* Milwaukee, WI: Trade Press Publication Co. (Monthly)
   Special issues include:
   January—Annual Buyers Guide
   August—What's New Issue (review of new products)

237. *Soap/Cosmetics/Chemical Specialties.* New York: MacNair-Dorland Co., Inc. (Monthly)

DRUGS, PHARMACEUTICALS

This listing presents information on prescription or over-the-counter drugs and biologicals such as blood and blood products, diagnostics, and vaccines. Biotechnology products include monoclonal antibodies for treating diseases, genetic engineering products, and interferon. Sources on the health care industry are also included.

Biographical Directories

238. *Who's Who in Health Care.* 2d ed. Edited by E. Sainer. Rockville, MD: Aspen Systems Corp., 1982. 612 p.
   Biographical information on leaders in the health field.

Encyclopedias, Dictionaries, and Handbooks

239. *American Hospital Association Guide to the Health Care Field.* Chicago: American Hospital Association, 1986. 650 p. (Annual)
   Directory of hospitals, multihospital systems, and health-related organizations.

240. *Dictionary of Biotechnology.* New York: Elsevier Science Publishing Co., Inc., 1986. 320 p.
   This dictionary defines the terminology, procedures, and processes that occur in biotechnology.

241. *Drug and Cosmetic Catalog, 1987.* New York: Harcourt Brace Jovanovich, Inc. (Annual)
   Lists drug and cosmetic manufacturers.

242. *Drug Topics Red Book, 1987.* Oradell, NJ: Medical Economics Co. (Annual)
   Manufacturers of prescription drugs and patent medicines and other drug store items.

243. *Genetic Engineering and Biotechnology Related Firms Worldwide Directory, 1987/88.* 6th ed. Kingston, NJ: Sitting & Noyes, 1986. 900 p.
   Information on some 3,000 companies worldwide involved in the biotechnology business.

244. *Medical and Healthcare Marketplace Guide.* 4th ed. Miami, FL: International Bio-Medical Information Service, Inc, 1986. 1,272 p.
   Covers public and private companies involved in the medical and health care industry, including pharmaceutical manufacturers and firms engaged in biochemicals and biotechnology in the United States.

245. *The Merck Index; An Encyclopedia of Chemicals, Drugs, and Biologicals.* 10th ed. Edited by Martha Windholz. Rahway, NJ: Merck & Co., Inc., 1983. 1,463 p.
   Covers organic and inorganic chemicals and drugs marketed worldwide.

246. *National Association of Chain Drug Stores—Membership Directory.* Alexandria, VA: National Association of Chain Drug Stores. (Annual)
   Chain drug retailers and individual pharmacies.

247. *National Wholesale Druggist's Association—Membership and Executive Directory, 1987.* Alexandria, VA: National Wholesale Druggists Association. (Annual)

248. *Physicians' Desk Reference, 1986.* 40th ed. Oradell, NJ: Medical Economics, Inc. (Annual)
   Manufacturers of prescribed drug products. Has a product information section where information on dosage, side effects, etc., of each drug can be found.

249. *USAN and the USP Dictionary of Drug Names.* Edited by Mary C. Griffiths. Rockville, MD: USP Convention, Inc., 1987. 690 p.
   The authoritative list of established names for drugs in the U.S. Names of drugs are adopted by the U.S. Adopted Names Council, which is cosponsored by the American Medical Association, the American Pharmaceutical Association, and the Pharmacopeial Convention.

250. *World Directory of Pharmaceutical Manufacturers.* London: IMSWORLD Publications, Ltd. (Biennial)
   Lists leading drug companies in major markets.

Indexes, Abstracts, and Continuing Services

251. *BIOSCAN; The Biotechnology Corporate Directory Service.* Phoenix, AZ: Oryx Press. (Bimonthly)
   Data on companies, investments, research, and developments in biotechnology. Compiled by the Cetus Corp. Information Services.

252. *International Pharmaceutical Abstracts.* Bethesda, MD: American Society of Hospital Pharmacists. (Semimonthly)

Newspapers, Journals, and Newsletters

253. *American Druggist.* New York: The Hearst Corp. (Monthly)
   The May issue features a prescription survey.

254. *Biomedical Products.* Dover, NJ: Gordon Publications, Inc. (Monthly)

255. *Drug and Cosmetic Industry.* Cleveland, OH: Harcourt Brace Jovanovich, Inc. (Monthly)

256. *Drugstore News.* New York: Lebhar-Friedman, Inc. (25/year)
   Special issues include:
   March—Top 100 Ethical Over-the-Counter Drugs
   April—Top 100 Pharmaceuticals

257. *Drug Topics.* Oradell, NJ: Medical Economics Co. (Semimonthly)
   Special issues include:
   April—Prescription Survey
   December—Business Outlook.

258. *The Green Sheet; Weekly Pharmacy Reports.* Chevy Chase, MD: F-D-C Reports, Inc. (Weekly)
   Gives news on the introduction and pricing of new pharmaceuticals and regulatory activity affecting pharmacists.

259. *Health Industry Today.* Springfield, NJ: Cassak Publications Co., Inc. (Monthly)

34 *Chemical Industries: An Information Sourcebook*

260. *Marketletter*. London: IMSWORLD Publications, Ltd. (Weekly)

261. *Medical Device & Diagnostic Industry*. Santa Monica, CA: Canon Communications, Inc. (Monthly)

262. *Medical Devices, Diagnostics and Instrumentation Reports: The Gray Sheet*. Chevy Chase, MD: F-D-C Reports, Inc. (Weekly)
Provides coverage of the medical devices, diagnostics and instrumentation industries, and includes regulatory and congressional activities.

263. *Medical Marketing & Media*. Ridgefield, CT: CPS Communications Inc. (Monthly)

264. *Medical World News*. Houston, TX: HEI Publishing, Inc. (Biweekly)

265. *Modern Healthcare*. Chicago: Crain Communications, Inc. (Monthly)
Special issues include:
January—Economic Outlook
May—Multi-unit Providers Survey; Rankings and Analyses of Health Care Providers.

266. *Pharmacy Times*. Port Washington, NY: Romaine Pierson Publishers, Inc. (Monthly)

267. *PMA Newsletter*. Washington, DC: Pharmaceutical Manufacturers Association. (Weekly)

268. *The Pink Sheet*. Chevy Chase, MD: F-D-C Reports, Inc. (Weekly)
Provides information on prescription and over-the-counter drugs.

269. *SCRIP—World Pharmaceutical New*. Surrey, England: PJB Publications, Ltd. (2/week)

270. *Washington Drug Letter*. Arlington, VA: Washington Business Information, Inc. (Weekly)

U.S. Government Publications

271. U.S. Department of Commerce. International Trade Administration. *Medical Equipment and Supplies Worldwide*. Washington, DC: Government Printing Office, 1984. 170 p.

272. U.S. Department of Commerce. International Trade Administration. Industry Analysis Division. *A Competitive Assessment of the U.S. Pharmaceutical Industry*. Washington, DC: Government Printing Office, 1985. 108 p.

ELECTRONICS, ELECTRICAL

Items displayed in this section contain information on computers, communications apparatus, analytical instruments, appliances, and the electrical utility industry.

Biographical Directories

273. *American Electronics Association; Membership Directory, 1987.* Palo Alto, CA: American Electronics Association. (Annual)
   Lists U.S. electronics and high technology companies; names and titles of key executives.

274. *Who's Who in Electronics, 1987.* Twinsburg, OH: Harris Publishing Co. (Annual)
   Manufacturers, manufacturers representatives, and distributors.

Encyclopedias, Dictionaries, and Handbooks

275. *Electronic Business 200 Issue, 1986.* Boston: Reed Holdings, Inc., Cahners Publishing Co. (Annual)
   100 companies with greatest revenues from sales of electronic products.

276. *Electronic Market Data Book.* Washington, DC: Electronic Industries Association, 1986. 158 p.
   Marketing statistics on the electronic industries.

277. *Electronic Marketing Directory, 1986.* New York: Dun & Bradstreet, Inc. (Annual)
   Electronic component manufacturers listed with Dun & Bradstreet credit reporting service.

278. *Electronic News Financial Fact Book & Directory, 1986.* New York: Fairchild Books. 500 p. (Annual)
   Corporate profiles, subsidiary and division details of publicly held electronics and electronics affiliated companies in the United States and Canada.

279. *Electronics Buyers Guide. 1986/87.* New York: McGraw-Hill, Inc. (Annual)
   Covers worldwide companies who manufacture electronic components, equipment, and allied products.

280. *Encyclopedia of Electronics.* Blue Ridge Summit, PA: Tab Books, Inc., 1985. 1,024 p.

36 *Chemical Industries: An Information Sourcebook*

281. *The Fiber Optics Sourcebook.* 2d. ed. Potomac, MD: Phillips Publishing Inc., 1987. 262 p.
 Information on fiber optic companies, networks, manufacturers and distributors, and business and technical services.

282. *International Fiber Optics and Communications; Handbook and Buyer's Guide Issue, 1987.* Boston: Information Gatekeepers, Inc., 1986. (Annual)
 Lists manufacturers and suppliers of fiber optics and communication products.

283. *Statistical Panorama, 1987.* Troy, MI: Business News Publishing Co. (Annual) 130 p.
 Forecasts industry trends and statistics on air conditioning, heating, and refrigeration markets.

284. *Statistical Yearbook of the Electric Utility Industry, 1987.* Washington, DC: Edison Electric Institute. (Annual) 110 p.

285. *World Electronics Yearbook, 1987.* Luton Beds, England: Benn Electronics Publications, Ltd., 1986. 452 p. 2 vols.
 Presents data from a survey of thirty countries which represent 95 percent of electronics industry markets. Includes statistics on products.

Indexes, Abstracts, and Continuing Services

286. *Electrical & Electronics Abstracts.* Piscataway, NJ: INSPEC, I.E.E. (Monthly)

Newspapers, Journals, and Newsletters

287. *Air Conditioning, Heating & Refrigeration News.* Troy, MI: Business News Publishing Co. (Weekly)
 The January issue features a directory of manufacturers, product listings, and trade names.

288. *Appliance.* Elmhurst, IL: Dana Chase Publications, Inc. (Monthly)
 Special issues include:
 September—Portrait of the U.S. Appliance Industry
 December—New Products/Annual Consumer Electronics Report.

289. *Appliance Manufacturer.* Denver, CO: Cahners Publishing Co. (Monthly)
 The January issue features an annual profile including who's who and market shares.

290. *Cellular Business.* Overland, KS: Intertec Publishing Corp.. (Monthly)

Categorical Arrangement 37

291. *Communications Week.* Manhasset, NY: CMP Publications, Inc. (Weekly)

292. *Consumer Electronics.* New York: CES Publishing Corp. (Monthly)
Special issues include:
January—Buyers Guide
March—Annual Statistical Forecast.

293. *Data Communications.* New York: McGraw-Hill Publications. (Monthly)
The September issue features DC 50—the fifty largest companies in the data communications business.

294. *Dealerscope Merchandising.* Philadelphia, PA: North American Publishing Co. (Monthly)
Special issues include:
April—Annual Statistical and Marketing Report; Spotlight on Major Appliances
May—Distributor and Manufacturers Rep Directory.

295. *Defense Electronics.* Palo Alto, CA: E. W. Communications, Inc. (Monthly)
Publishes a Marketing Directory and Buyers Guide in February or March.

296. *EE—Electronic/Electrical Product News.* White Plains, NY: Sutton Publishing Co. (Monthly)

297. *Electric Light & Power.* Barrington, IL: Technical Publishing Co. (Monthly)
Special issues include:
June—Top 100 Electric Utilities—sales and financial performance
July—Top 50 Publicly Owned Electric Utilities
August—Top 100 Electric Utilities—operating performances.

298. *Electric Utility Week.* New York: McGraw-Hill Publications, Inc.

299. *Electrical World.* New York: McGraw-Hill, Inc. (Monthly)
Special issues include:
March/April—Annual Statistical Report
September—Electric Utility Industry Forecast

300. *Electronic Business.* Denver, CO: Cahners Publishing Co. (Monthly)
Special issues include:
January—Annual Forecast
February/August—Top 100 Electronic Manufacturers

301. *Electronic Business Forecast.* San Jose, CA: Cahners Publishing Co. (Semimonthly)

302. *Electronic Chemicals & Materials News.* Maplewood, NJ: CTB International Publishing Co. (Monthly)

303. *Electronic Chemicals News.* New York: Chemical Week, McGraw-Hill, Inc. (Biweekly)

304. *Electronic Market Trends.* Washington, DC: Electronic Industries Association. (Monthly)

305. *Electronic Materials Report.* Los Altos, CA: Rose Associates. (Monthly)

306. *Electronic News.* New York: Fairchild Publications, Inc. (Weekly)

307. *Electronic Packaging & Production.* Newton, MA: Reed Publishing, U.S.A., Cahners Magazine Division. (Monthly)

308. *Electronic Products Magazine.* Garden City, NY: United Technical Publications, Inc., Hearst Business Communications, Inc. (15/year)

309. *Electronics.* New York: McGraw-Hill, Inc. (Weekly)
 The January issue features World Market Forecast.

310. *Fiber Optics News.* Bethesda, MD: Phillips Publishers, Inc. (Weekly)

311. *Fiberoptics Report.* Littleton, MA: Advanced Technology Publications. (22/year)
 The January issue features Annual Economic Review and Outlook.

312. *High Technology.* Boston: High Technology Publishing Corp. (Monthly)

313. *Hybrid Circuit Technology.* Libertyville, IL: Lake Publishing Corp. (Monthly)

314. *Laser Report.* Littleton, MA: Advanced Technology Publications. (Monthly)
 The January issue features Annual Economic Review and Outlook.

315. *Lasers & Applications.* Torrance, CA: High Tech Publications. (Monthly)

316. *Power.* New York: McGraw-Hill Publications Co. (Monthly)

317. *Printed Circuit Fabrication.* Alpharetta, GA: PMS Industries. (Monthly)

318. *Public Power.* Washington, DC: American Public Power Association. (6/year)
 The January/February issue features Directory of Local Publicly Owned Electric Utilities.

319. *Public Utilities Fortnightly.* Arlington, VA: Public Utilities Reports, Inc. (26/year)

320. *SIA Circuit Newsletter.* Cupertino, CA: Semiconductor Industry Association. (Monthly)

321. *Solid State Technology.* Port Washington, NY: Technical Publishing Co. (Monthly)

322. *Telephony.* Chicago: Telephony Publishing Corp. (Weekly)
   The January issue features Review and Forecast.

U.S. Government Publications

323. U.S. Department of Energy. Energy Information Administration. *Steam-Electric Plant Construction Cost and Annual Production Expenses.* Washington, DC: Government Printing Office. (Annual)

324. U.S. Department of Energy. Energy Information Administration. Office of Coal, Nuclear, Electric, and Alternate Fuels. *Annual Outlook for United States Electric Power.* Washington, DC: Government Printing Office, 1986. 73 p.

325. U.S. Department of Energy. Energy Information Administration, Office of Coal, Nuclear, Electric, and Alternate Fuels. *Financial Statistics of Selected Electric Utilities, 1984.* Washington, DC: Government Printing Office, 1986. 806 p.

326. U.S. Department of Energy. Office of Utility Project Operations. *Inventory of Power Plants in the United States, 1985.* Washington, DC: Government Printing Office, 1986. 306 p.

MINING, MINERALS, METALS

This large category includes references related to the activities and products of the mining, minerals, and metals industries. These include metallic and nonmetallic minerals, industrial gases, aluminum, magnesium, mercury, brine, bromine, calcium compounds, carbide, cobalt, copper, flouride, gold, hydrochloric acid, hydrogen peroxide, iodine, iron, lithium, molybdenum, nickel, phosphorous, potash, potassium, ribidum, salt, silica, sodium, strontium, sulfur, tin, titanium, uranium, zinc chloride, stone, clay, glass, concrete, ceramics, coal, and steel.

Electronic Retrieval Systems

327. FAST TRACK. New York: Metal Bulletin, Inc.
Provides daily information on mineral commodities, price series, company events, tender awards, and industrial contracts.

Encyclopedias, Dictionaries, and Handbooks

328. *Concrete Admixtures Handbook; Properties, Science and Technology.* Edited by V. S. Ramachandran. Park Ridge, NJ: Noyes Data Corp., 1984. 626 p.
A basic treatise covering cement science, properties, patents. Various sections authored by known experts in the field.

329. *Directory of Iron and Steel Works of the U.S. and Canada.* 36th ed. Washington, DC: American Iron and Steel Institute, 1984. 359 p.

330. *Engineering and Mining Journal International Directory of Mining and Mineral Processing Operations, 1986.* New York: McGraw-Hill, Inc. 600 p. (Annual)
Mines and plants producing metals and nonmetallic minerals.

331. *Financial Times International Yearbook—Mining.* Harlow, Essex, England: Longman Group, Ltd. (Irregular)
Companies involved in the mining, production, and distribution of minerals and ores worldwide.

332. *Glass Factory Directory, 1986.* Pittsburgh, PA: The National Glass Budget. 140 p. (Annual)
U.S. and Canadian glass manufacturers and plants.

333. *The Glass Industry Directory, 1987.* New York: Astlee Publishing Co. 230 p. (Annual)
Lists worldwide glass manufacturers and related firms.

334. *Industrial Minerals and Rocks.* 5th ed. New York: Society of Mining Engineers, American Institute of Mining, Metallurgical, and Petroleum Engineers, 1983. 2 vols. 1,446 p.
This source has information on industrial minerals grouped by uses, including the chemical industry and a section giving detail on individual commodities and other sources of information for industrial minerals.

335. *Industrial Minerals Directory; World Guide to Producers and Processors, 1986.* 1st ed. New York: Metal Bulletin, Inc. 704 p. (Annual)
A worldwide list of mineral producers and processors giving the name, address, telephone/telex/facsimile numbers, location, and description of processing facilities and minerals produced.

336. *Iron Ore Databook.* 1st ed. New York: Metal Bulletin, Inc., 1986. 156 p.
Contains information on international iron ore, mines, traders, representatives, import/export data, production statistics, key prices, and projections.

337. *Keystone Coal Industry Manual, 1986.* New York: McGraw-Hill, Inc. 1,306 p. (Annual)
Coal companies and mines, coke plants, coal cleaning plants, etc. Includes list of leading coal mining companies and mines.

338. *Metal Statistics, 1986.* New York: American Metal Market. (Annual)

339. *Mineral Producers and Processors Directory, 1985.* Morgantown, WV: West Virginia Geological and Economic Survey, 1986. (Annual)

Indexes, Abstracts, and Continuing Services

340. *Ceramic Abstracts.* Columbus, OH: American Ceramic Society. 1922–. (Bimonthly)
Covers worldwide literature on scientific, engineering, and commercial aspects of ceramics and related materials. Also available on PERGAMON INFOLINE.

341. *Metals Abstracts.* Metals Park, OH: American Society for Metals. 1968–. (Monthly)

342. *Mineralogical Abstracts.* London: Mineralogical Society. 1959–. (Quarterly)

343. *World Aluminum Abstracts.* Washington, DC: Aluminum Association, Inc. 1968–. (Monthly)

Newspapers, Journals, and Newsletters

344. *American Glass Review.* Clifton, NJ: Ebel-Doctorow Publications, Inc. (Monthly)
The February issue features the Glass Factory Directory, a listing of U.S. manufacturers and suppliers.

345. *American Metal Market.* New York: Fairchild Publications, Inc. (Daily)
Special issues deal with specific commodities. Regular features of each issue are price sections.

346. *Brick and Clay Record.* Des Plaines, IL: Cahners Publishing Co. (Monthly)
Special issues include:
June—Annual Industry Forecast
December—International Market Outlook.

347. *Ceramic Industry.* Denver, CO: Cahners Publishing Co. (Monthly)
Special issues include:
June—Annual Review and Forecast
August—Ceramic Industry Giants
October—Ceramic Data Book.

348. *Coal Age.* New York: McGraw-Hill, Inc. (Monthly)
Publishes a Review and Outlook issue in February.

349. *Coal Outlook.* Arlington, VA: Pasha Publications. (Weekly)
Features coal prices and production.

350. *Coal Week.* New York: McGraw-Hill, Inc.
Covers U.S. steam coal prices, markets, politics, and economics.

351. *Compressed Air.* Washington, NJ: Compressed Air Magazine, Co. (Monthly)

352. *Concrete Products.* Chicago: Maclean-Hunter Publishing Corp. (Monthly)

353. *Engineering & Mining Journal.* New York: McGraw-Hill Publications Co. (Monthly)
Publishes a Mineral Commodities Survey with production statistics in March.

354. *Engineering News Record.* New York: McGraw-Hill, Inc. (Weekly)
Special issues include:
January—Annual Report & Forecast
April—Top 400 Contractors
May—Top 500 Design Firms.

355. *Glass Industry.* New York: Ashlee Publishing Co., Inc. (Monthly)

356. *Industrial Minerals.* New York: Metal Bulletin, Inc. (Monthly)

357. *Iron Age.* Radnor, PA: Chilton Co. (24/year)
Has a regular section on materials prices. Special issues include:
January—Forecasts and Statistical Summaries
May—Steel Industy Outlook.

358. *Metal Bulletin.* New York: Metal Bulletin, Inc. (2/week)

359. *Metals Week.* New York: McGraw-Hill Publications Co. (Weekly)

360. *Metals Week Insider Report (Telex Service).* New York: McGraw-Hill Publications Co. (Daily)

361. *Mining Magazine.* London: Mining Journal, Ltd. (Monthly)

362. *Modern Metals.* Chicago: Modern Metals Publishing Co. (Monthly)

363. *Pit & Quarry.* Cleveland, OH: Harcourt Brace Jovanovich (Monthly)
   Features Outlook/Review in January.

364. *Rock Products.* Chicago: MacLean Hunter Publishing Co. (Monthly)
   In December publishes a forecast issue.

365. *Skillings Mining Review.* Duluth, MN: Skillings Mining Review, Inc. (Weekly)

366. *33 Metal Producing.* New York: McGraw-Hill Publications Co. (Monthly)

367. *U.S. Glass, Metal & Glazing.* Memphis, TN: U.S. Glass Publications, Inc. (6/year)
   Special issues include:
   January/February—Outlook & Forecast Issue
   August—Sealants and Glazing Systems
   November—Buyers Guide Issue

368. *World Coal.* San Francisco, CA: Miller Freeman Publications. (6/year)

369. *World Mining.* New York: Technical Publishing Co., Dun & Bradstreet. (Bimonthly)
   In August publishes *Yearbook*, a catalog of new product literature, mining trends, production statistics by country, equipment, and manufacturers directory.

U.S. Government Publications

370. U.S. Department of Commerce. Bureau of the Census. *Census of Mineral Industries, 1982.* Washington, DC: Government Printing Office, 1985.
   This economic census is conducted every five years and contains data on establishments engaged in the extraction of minerals in the United States. The series includes information on industries, geographic areas, and subjects.

371. U.S. Department of Energy. Energy Information Administration. *Domestic Uranium Mining and Milling Industry (1985 Viability Assessment)*. Washington, DC: Government Printing Office, 1986. 156 p.
   This report is an annual assessment of the domestic uranium mining and milling industry and also gives projections through the year 2000.

372. U.S. Department of Energy. Energy Information Administration. *Uranium Industry Annual 1985*. Washington, DC: Government Printing Office, 1986. 156 p.
   A statistical profile of the U.S. uranium industry is presented in this report and includes data on uranium raw materials, prices, imports and exports, and other marketing activities.

373. U.S. Department of Energy. Energy Information Administration. *Weekly Coal Production*. Washington, DC: Government Printing Office.
   Provides current information on coal supply and demand, production, consumption, prices, stocks, and exports.

374. U.S. Department of Energy. Energy Information Administration. Office of Energy Markets and End Use. *Monthly Energy Review*. Washington, DC: Government Printing Office. (Monthly)
   Presents current data on production, consumption, stocks, imports, exports, and prices of the principal energy commodities in the U.S.

375. U.S. Department of the Interior. Bureau of Mines. *Mineral Commodity Summaries, 1987*. Washington, DC: Government Printing Office. (Annual)
   Current summaries of eighty-seven nonfuel mineral commodities; the data following the format of the chapters in *Mineral Facts and Problems* (see entry 388).

376. U.S. Department of the Interior. Bureau of Mines. *Mineral Facts and Problems*. Bureau of Mines Bulletin 675. Washington, DC: Government Printing Office, 1985. 964 p.
   Published every five years, this is a standard reference on mineral commodities in the U.S. and abroad.

377. U.S. Department of the Interior. Bureau of Mines. *Mineral Industry Surveys*. Washington, DC: Government Printing Office.
   Contains domestic and foreign statistical and economic data on various mineral commodities. Reports are issued weekly, monthly, quarterly, or annually.

378. U.S. Department of the Interior. Bureau of Mines. *The Mineral Position of the United States: The Past Fifteen Years, 1971-1985*. Washington, DC: Government Printing Office, 1986. 45 p.

379. U.S. Department of the Interior. Bureau of Mines. *Minerals Yearbook.* Washington, DC: Government Printing Office. 3 vols. (Annual)
> Vol. I—*Metals and Minerals*, 1985. Contains chapters on all metallic and nonmetallic mineral commodities important to the U.S. economy.
> Vol. II—*Area Reports: Domestic*, 1986. A list of principal producers of iron ore, cement, lime, gypsum, clays, sand and gravel, phosphatic rock, and other nonfuel minerals in the U.S.
> Vol. III—*Area Reports: International*, 1986. Mineral data on foreign countries and discussions of the importance of minerals to the economies of these nations.

380. U.S. Department of the Interior. Bureau of Mines. *The 1985 Annual Report of the Secretary of the Interior Describing the Nonfuel Mineral Industry; Supply, Demand and Outlook.* Washington, DC: Government Printing Office.

PAINTS, VARNISHES, LACQUERS, ENAMELS, COATINGS, ADHESIVES AND SEALANTS

Paints and coatings are made from many raw materials, and markets for these products fall into three categories: architectural, product, and special purpose. Other coating products include varnishes, lacquers, and enamels. Adhesives and sealants are also formulated products and are used in a variety of applications in the construction, motor vehicle assembly, furniture manufacturing, and appliance and electronic equipment manufacturing industries.

Encyclopedias, Dictionaries, and Handbooks

381. *Adhesives Age Directory.* 18th ed. Atlanta, GA: Communication Channels, 1986. 218 p. (Annual)
> Lists manufacturers, products, suppliers, consultants.

382. *Adhesives Technology Handbook.* Edited by Arthur H. Landrock. Park Ridge, NJ: Noyes Data Corp., 1985. 444 p.
> Covers applications, properties, design, and preparation techniques.

383. *Federation of Societies for Coatings Technology—Yearbook and Membership Directory,* 1987. Philadelphia, PA: Federation of Societies for Coatings Technology. 328 p. (Annual)

384. *Paint Red Book.* 19th ed. Atlanta, GA: Communication Channels, Inc., 1987. 288 p. (Annual)
> Lists paint and coating manufacturers in the U.S., Canada, and Puerto Rico: brand names, raw material suppliers, machinery and equipment makers, consulting firms, and testing labs.

385. *Painting and Wallcovering Contractor—PDCA Yearbook Issue.* Falls Church, VA: Painting & Decorating Contractors of America. (Annual)
> Contractors engaged in painting, decorating, drywall, wallcoverings, and special coatings application.

Indexes, Abstracts, and Continuing Services

386. *World Surface Coatings.* Oxford, England: Paint Research Association, Pergamon Press. (Monthly)
> Covers worldwide literature on all aspects of the paint and surface coatings industries. Also available on PERGAMON INFOLINE.

Newspapers, Journals, and Newsletters

387. *Adhesives Age.* Atlanta, GA: Communication Channels. (Monthly)

388. *Adhesives and Sealants Newsletter.* Berkeley Heights, NJ: Fred Keimel. (Monthly)

389. *American Paint & Coatings Journal.* St. Louis, MO: American Paint Journal Co. (Weekly)
> Contains current prices of materials.

390. *Industrial Finishing.* Wheaton, IL: Hitchcock Publishing Co. (Monthly)
> The September issue includes New Product and Literature Review and Directory Issue.

391. *Journal of Coatings Technology.* Philadelphia, PA: Federation of Societies for Coatings Technology. (Monthly)

392. *Modern Paint & Coatings.* Atlanta, GA: Communication Channels, Inc. (Monthly)
> Each issue includes "Business Activity Indicators," which give prices of coatings and paints. The January issue features Annual Review and Forecast.

393. *Paint and Coatings Industry.* Canoga Park, CA: Western Trade Publishing Co. (Monthly)

394. *Products Finishing Magazine.* Cincinnati, OH: Gardner Publications, Inc. (Monthly)
> The September issue features the Products Finishing Directory Issue.

PAPER AND ALLIED PRODUCTS

The pulp, paper, and paperbound industry includes products such as newsprint, printing papers, tissue and sanitary papers, fine papers, specialized industrial papers and linerboard, boxboard, foodboard, and construction board. This grouping also includes forest products such as lumber and plywood and paper mills, pulp mills, converted paper products, and paperboard packaging.

Biographical Directories

395. *Paper Industry Management Association—Membership Directory, 1986/87.* Arlington Heights, IL: Paper Industry Management Association. (Annual)
  Lists pulp, papermill, and paper converting production executives.

Encyclopedias, Dictionaries, and Handbooks

396. *Crow's Buyers and Sellers Guide of the Forest Products Industries, 1986.* Portland, OR: C. C. Crow Publications, Inc. (Annual)
  Manufacturers of lumber, plywood, and other forest products.

397. *Directory of the Forest Products Industry, 1986.* San Francisco, CA: Miller Freeman Publications, Inc. (Annual)
  Sawmills, plywood, board, and veneer mills in the United States.

398. *Lockwood's Directory of the Paper and Allied Trades, 1987.* New York: Vance Publishing Corp. (Annual)
  Directory of pulp and paper mills and converters.

399. *Paper Year Book, 1987.* 45th ed. Duluth, MN: Harcourt Brace Jovanovich, Inc. (Annual)
  List of manufacturers and suppliers of paper and paper-related products. Contents also include descriptions of major product lines, markets, recent trade developments, etc.

400. *Post's Pulp & Paper Directory, 1986.* San Francisco, CA: Miller Freeman Publications, Inc. (Annual)
  North American pulp and paper mills, converting plants, associations, schools and research facilities, suppliers of equipment, services and chemicals.

401. *Pulp & Paper Capacities Survey, 1984-1989.* New York: United Nations, FAO, 1985. 178 p.
  Contains the *Report on the FAO Survey of World Pulp and Paper Capacities 1983-1988* as well as updated statistical data for countries of Africa, North and Central America, South America, Asia, Europe, Oceania, and the Soviet Union.

48 Chemical Industries: An Information Sourcebook

402. *Pulp and Paper Dictionary.* Edited by John R. Lavigne. San Francisco, CA: Miller Freeman Publications, Inc., 1986. 370 p.

403. *Pulp & Paper—North American Industry Factbook, 1984-85.* San Francisco, CA: Miller Freeman Publications, Inc. (Biennial)
United States and Canadian pulp and paper companies with a total capacity of 1,000 tons or more.

Indexes, Abstracts, and Continuing Services

404. *Abstract Bulletin of the Institute of Paper Chemistry.* Appleton, WI: Institute of Paper Chemistry. (Monthly)

405. *Paper Industry News Digest.* Appleton, WI: Albany International Appleton Wire Division. (Biweekly)
Abstracts of articles on the paper industry from other publications.

406. *PIRA Abstracts.* Surrey, England: Paper & Board, Printing & Packaging Industries Research Association.
Covers the literature on all aspects of paper and board making, printing and packaging. Also available on PERGAMON INFOLINE.

407. *Pulp & Paper International Newswire.* San Francisco, CA: Miller Freeman Publications. (Weekly)
A newservice for the worldwide paper industry.

Newspapers, Journals, and Newsletters

408. *Boxboard Containers.* Chicago: Maclean-Hunter Publishing Corp. (Monthly)

409. *Building Supply and Home Centers.* Chicago: Cahners Publishing Co., Inc. (Monthly)

410. *Forest Industries.* San Francisco, CA: Forest Industries Circulation Department. (Monthly)

411. *Monthly Report on Pulp and Paper Mill Projects in the World.* Helsinki, Finland: Jaakko Poyry Oy. (11/year)

412. *Monthly Statistical Summary.* New York: American Paper Institute. (Monthly)
Provides pulp, paper, and paperboard statistics.

413. *Paper, Film & Foil Converter.* Chicago: Maclean-Hunter Publishing Corp. (Monthly.)
The June issue features the Buyers Guide & Directory.

414. *Paper, Paperboard & Wood Pulp Capacity.* New York: American Paper Institute. (Monthly, Annual)
   Publishes Statistics of Paper, Paperboard and Wood Pulp in September.

415. *Paper Trade Journal.* Lincolnshire, IL: Vance Publishing Corp. (Annual)
   Special issues include:
   June—Top 50 Paper Companies
   August—Chemical Survey of Pulp and Paper Industry
   December—Focus on Economics for Pulp and Paper.

416. *Paperboard Packaging.* New York: Magazines for Industry, Inc. (Monthly)
   Special issues include:
   January—Buyers Forecast
   August or September—Statistics

417. *Pulp & Paper.* San Francisco, CA: Miller Freeman Publications, Inc. (Monthly)
   Special issues include:
   May—Coating Annual/Register of Adhesives & Binders
   August—Worldwide Roundup
   October—Annual Chemicals Review
   November—Buyers Guide.

418. *Pulp & Paper Week.* San Francisco, CA: Miller Freeman Publications, Inc.

419. *Random Lengths.* Eugene, OR: Random Length Publications, Inc. (Weekly)

420. *Wood Products.* Lincolnshire, IL: Vance Publishing Corp. (Monthly)
   Special issues include:
   January—Outlook
   March—Reference Buying Guide

421. *World Wood.* San Francisco, CA: Miller Freeman Publications, Inc. (Monthly)
   The August issue features the World Wood Review, listing trade statistics and production figures.

U.S. Government Publications

422. U.S. Department of Commerce. Bureau of Industrial Economics. *Construction Review.* Washington, DC: Government Printing Office. (Bimonthly)

PETROCHEMICALS, ENERGY

The oil and gas industry fulfills most of the nation's needs for energy. Some of the major products are natural gas, motor gasoline, fuel oil, and jet fuel. Many products derived from petroleum are petrochemicals used to produce other chemical products such as plastics, synthetic rubber, synthetic fibers, surface active agents, plasticizers, solvents, nitrogenous fertilizers, and carbon black.

Biographical Directories

423. *Financial Times Who's Who in World Oil and Gas, 1982-83.* 7th ed. Philadelphia, PA: International Publications Service, Division of Taylor and Francis, 1982. 636 p.
   Senior executives, technologists, scientists, government representatives, and consultants in the petroleum and gas industry.

424. *International Who's Who in Energy and Nuclear Sciences.* Harlow, Essex, England: Longman Group, Ltd.; dist. by Gale Research Co., Detroit, MI, 1983. 531 p.
   Research chemists, physicists, and development engineers concerned with the generation, storage, and use of energy.

425. *The International Who's Who of the Arab World.* 3d ed. Houston, TX: Pennwell Publishing Co, 1984. 600 p.

426. *Pipeline Digest Who's Who in Pipelining, 1983.* Houston, TX: Universal News, Inc. (Annual)
   Pipeline contractors and subcontractors, pipeline operating companies, including those for crude oil, natural gas, petrochemicals, coal slurry, petroleum products, and liquid natural gas.

427. *Who's Who in World Petrochemicals.* Houston, TX: DeWitt & Co., Inc., 1986. 247 p.
   A listing of people in the international petrochemical industry.

Electronic Retrieval Systems

428. APILIT. New York: American Petroleum Institute. (Monthly)
   Information on petroleum products and technology and the economics of chemicals used in the oilfield. Also available on SDC Information Services, Inc.

429. EPUB Electronic Publication System. Washington, DC: Energy Information Administration.
   Selected Weekly Petroleum Status Report and Petroleum Supply Monthly Statistics; updated weekly and monthly.

430. OIL AND GAS JOURNAL ENERGY DATABASE. Tulsa, OK: Pennwell Publishing Co. (Daily)
> Statistical information about the oil and gas industry. Time series covers drilling and exploration, production, reserves, refining, imports and exports, demand and consumption, transportation, and prices.

431. PETROLEUM/ENERGY BUSINESS NEWS. New York: American Petroleum Institute. 1975–. (Weekly)
> Political, social, and economic news related to the energy industries. Available on SDC Information Services, Inc.

432. TULSA. Tulsa, OK: Petroleum Abstracts, University of Tulsa. (Weekly)
> Covers technical literature and patents on the exploration and production of gas and oil, drilling and mineral commodities. Also available on SDC Information Services, Inc.

Encyclopedias, Dictionaries, and Handbooks

433. *Basic Petroleum Data Book, 1987.* Washington, DC: American Petroleum Institute. (3/year)
> Data include domestic and world statistical background information on energy, reserves, exploration and drilling, production, finance, prices, demand, refining, imports and exports, natural gas, and the Organization of Petroleum Exporting Countries.

434. *Desk & Derrick Standard Oil Abbreviator.* 3d ed. Compiled by the Association of Desk & Derrick Club. Houston, TX: Pennwell Publishing Co., 1986. 318 p.

435. *Gas Facts, 1985.* Arlington, VA: The American Gas Association. (Annual)

436. *Gulf Coast Oil Directory, 1987.* Houston, TX: Resource Publications, Inc. (Annual)

437. *Illustrated Petroleum Reference Dictionary.* 3d ed. Edited by Robert D. Langenkamp. Houston, TX: Pennwell Publishing Co., 1985. 696 p.

438. *International Petroleum Encyclopedia, 1986.* Tulsa, OK: Pennwell Publishing Co. (Annual)
> Information on refining and pipeline developments, offshore activities, OPEC, and price trends.

439. *Oil & Gas Journal Data Book, 1986.* Tulsa, OK: Pennwell Publishing Co. (Annual)

440. *Oil and Gas Stocks Handbook.* New York: Standard & Poor's Corp. (Semiannual)

441. *Oil Industry Outlook, 1987-1991.* Tulsa, OK: Pennwell Publishing Co., 1986. 140 p.
Contains information on the current state and future of the industry.

442. *Oil Industry, U.S.A.* New York: Oil Daily, Inc. (Annual)

443. *U.S.A. Oil Industry Directory.* 25th ed., Tulsa, OK: Pennwell Publishing Co., 1986. 700 p. (Annual)
More than 5,000 companies listed, including major integrated companies, fund companies, retail and wholesale marketers, crude oil traders, state and federal agencies, and petroleum related associations. Also features the *Oil & Gas Journal* 400 report.

444. *Worldwide Petrochemical Directory.* Tulsa, OK: Pennwell Publishing Co., 1986. 280 p.
Information on worldwide companies involved in the petrochemical industry. Also contains the *Oil & Gas Journal's Worldwide Petrochemical Survey,* which details current petrochemical plants, feedstocks, products, capacities, and a petrochemical plant construction survey.

Indexes, Abstracts, and Continuing Services

445. *Diesel Fuel Oils, 1985.* By Ella Mae Shelton and Cheryl L. Dickson. Bartlesville, OK: National Institute for Petroleum and Energy Research, October 1985. (Annual)
Results of an analysis of diesel fuel oil produced in U.S. refineries.

446. *Heating Oils, 1986.* By Cheryl L. Dickson and Paul W. Woodward. Bartlesville, OK: National Institute for Petroleum and Energy Research, August 1986. (Annual)
This report contains information on heating oils manufactured in the U.S.

447. *Motor Gasolines, Summer, 1985.* By Cheryl L. Dickson and Paul W. Woodward. Bartlesville, OK: National Institute for Petroleum and Energy Research, June 1986. (Annual)
Analytical data for motor gasoline and motor gasoline/alcohol blend samples derived from the gasolines of sixty-two marketers.

448. *Petroleum Abstracts.* Tulsa, OK: University of Tulsa, Information Services Division. 1961–. (Weekly)
Worldwide coverage of literature relating to exploration, development, and production of oil and natural gas.

449. *Petroleum/Energy Business News Index.* New York: American Petroleum Institute. 1975–. (Monthly)

Newspapers, Journals, and Newsletters

450. *C4 Market Report.* Houston, TX: Chemical Market Associates, Inc. (Semimonthly)
Provides statistics on prices, exports, and imports.

451. *C4 Monitor.* Montclair, NJ: CTC International. (Monthly)

452. *Ethylene Service Newsletter.* Houston, TX: DeWitt & Co., Inc. (Biweekly)
Provides ethylene and derivatives prices.

453. *Fiber Intermediates Market Report.* Houston, TX: Chemical Market Associates, Inc. (Semimonthly)
Provides product prices of toluene, paraxylenes, ethylene glycol, benzene, cumene, phenol, acrylonitrile, and nylon.

454. *Hydrocarbon Processing.* Houston, TX: Gulf Publishing Co. (Monthly)
Special issues include:
February/June—HPI Construction Boxscore
April—Gas Processing Report.

455. *Inside FERC'S Gas Market Report.* New York: McGraw-Hill Publications, Inc. (Biweekly)

456. *International Petroleum Finance.* New York: Petroleum Analysis, Ltd. (Semimonthly)

457. *Monomers Market Report.* Houston, TX: Chemical Market Associates, Inc. (Semimonthly)
Publishes news items and spot and contract prices in the U.S., Western Europe, and Japan.

458. *Monthly Statistical Bulletin.* Washington, DC: American Petroleum Institute.

459. *National Petroleum News.* Des Plaines, IL. (Monthly)
Special issues include:
March—Buyers Guide Issue
June—Fact Book Issue.

460. *Natural Gas Liquids Update.* Houston, TX: Resource Planning Consultants, Inc. (Monthly)
Published as a supplement to a quarterly multiclient study *Natural Gas Liquids Outlook for Supply, Demand and Pricing.*

461. *Oil & Gas Journal.* Tulsa, OK: Pennwell Publishing Co. (Weekly)
Special issues include:
January—Pipeline Construction Survey/Forecast/Review
February—Capital Spending Survey
March or April—Annual Refining/Petrochemical/Production

May—Offshore Report
June—Exploration Development
July—Annual Gas Processing
August—Annual Pipeline Number
September—Annual Drilling Issue, Ethylene Report
November—Pipeline Economics
December—Worldwide Report

462. *Oil Industry Comparative Appraisals.* Greenwich, CT: John S. Herold, Inc. (Monthly)

463. *Olefins Market Letter.* Houston, TX: International PC, Inc. (Weekly)

464. *Petrochemical News.* Stamford, CT: William F. Bland. (Weekly)
Worldwide coverage of new business ventures, new plants and plant expansions, corporate changes, mergers and acquisitions, prices and markets, legislation, labor, etc.

465. *Petroleum Economist.* London: Petroleum Press Bureau, Ltd. (Monthly)

466. *Pipe Line Industry.* Houston, TX: Gulf Publishing Co. (Monthly)
January—Construction Report and Forecast

467. *Pipeline & Gas Journal.* Dallas, TX: Harcourt Brace Jovanovich, Inc. (14/year)
Special issues include:
January—Construction Forecast
August—P & GJ 500—Top Companies.

468. *Platt's Oilgram & Price Report.* New York: McGraw-Hill Publications, Inc. (Daily)

469. *Propylene Service Newsletter.* Houston, TX: DeWitt & Co., Inc. (Biweekly)
Publishes propylene and derivative prices.

470. *Toluene-Xylenes Newsletter.* Houston, TX: DeWitt & Co., Inc. (Weekly)
Features news, prices, production, consumption, stocks.

471. *U.S. Oil Week.* Alexandria, VA: Capitol Publications, Inc.
Reports on petroleum trends; supplement contains prices at key terminals.

472. *Weekly Propane Newsletter.* Arcadia, CA: Butane-Propane News.

473. *Weekly Statistical Bulletin.* Washington, DC: American Petroleum Institute.

474. *World Oil.* Houston, TX: World Oil. (Monthly)
Special issues include:
February—Outlook—Statistics and Forecast
June—Guide to Drilling, Completion, and Workover Fluids
July—Offshore Exploration and Production
August—International Outlook

U.S. Government Publications

475. U.S. Department of Energy. Energy Information Administration. *Natural Gas Annual, 1986.* Washington, DC: Government Printing Office. 288 p.
Information on the reserves, production, imports, exports, storage, consumption, and price of natural gas.

476. U.S. Department of Energy. Energy Information Administration. *Natural Gas Monthly.* Washington, DC: Government Printing Office.
This report provides monthly information at state and national levels on the supply and disposition of natural gas, including production, storage, imports, exports, and consumption.

477. U.S. Department of Energy. Energy Information Administration. *Petroleum Marketing Annual 1985.* Vol. 1. Washington, DC: Government Printing Office, 1986. 160 p.
Contains the annual data on crude oil and refined petroleum products. Gives prices, volume, and percentage statistics for 1985.

478. U.S. Department of Energy. Energy Information Administration. *Petroleum Marketing Annual 1985.* Vol. 2. Washington, DC: Government Printing Office, 1986. 412 p.
This is a compilation of the final 1985 monthly crude oil purchase price and percentage statistics and refined petroleum products sales statistics.

479. U.S. Department of Energy. Energy Information Administration. *Petroleum Supply Monthly.* Washington, DC: Government Printing Office.
Monthly statistics on petroleum supply, disposition, production, refinery operations, stocks, and transport.

480. U.S. Department of Energy. Energy Information Administration. *Weekly Petroleum Status Report.* Washington, DC: Government Printing Office.

481. U.S. Department of Energy. Energy Information Administration. Office of Energy Markets and End Use. *Future Markets and Petroleum Supply.* Washington, DC: Government Printing Office, 1986. 53 p.

482. U.S. Department of Energy. Energy Information Administration. Office of Energy Markets and End Use. *International Energy Outlook, 1985: With Projections to 1995.* Washington, DC: Government Printing Office, 1986. 66 p.

483. U.S. Department of Energy. Energy Information Administration. Office of Oil & Gas. *Petroleum Marketing Monthly.* Washington, DC: Government Printing Office.
   Price information and sales data on various petroleum products is presented in this publication.

484. U.S. House of Representatives. Committee on Energy and Commerce. *Compilation of Selected Energy Related Legislation.* Vol. 1, *Oil Gas, and Nonnuclear Fuels.* Washington, DC: Government Printing Office, 1985. 672 p.

PLASTICS MATERIALS, PACKAGING

The plastics materials industry is a large and growing industry and plastics will probably continue to displace such materials as wood, glass, and metal in many markets. Packaging is a large part of the consumption of plastic materials.

Biographical Directories

485. *Who's Who in Packaging.* Stamford CT: The Packaging Institute, U.S.A., 1978. 176 p.
   Lists packaging professionals in over fifty categories.

Electronic Retrieval Systems

486. Electronic Reporting System for the Monthly Statistical Report—RESINS. New York: Society for the Plastics Industry.

487. PLASPEC. New York: Plastics Technology, Bill Publications.
   An engineering and marketing databank which includes online pricing information (list and market), product and industry news, and suppliers directories.

Encyclopedias, Dictionaries, and Handbooks

488. *Encyclopedia of Packaging Technology.* New York: John Wiley & Sons, Inc., 1986. 800 p.
   Covers all aspects of packaging technology, from raw materials through distribution.

489. *Encyclopedia of Plastics, Polymers & Resins.* Compiled by Michael & Irene Ash. 3 vols. (Vol. 1—1981, 392 p.; Vol. 2—1982, 395 p.; Vol. 3—1983, 442 p.). Brookfield Center, CT: Society of Plastics Engineers.
   Gives practical information on plastic, polymer, and resin trademark products.

490. *Encyclopedia of PVC.* 2d ed. Edited by L. I. Nass and C. A. Heilberger. Brookfield Center, CT: Society of Plastics Engineers, 1985. 680 p.

491. *Facts & Figures of the U.S. Plastics Industry, 1986.* New York: Society of the Plastics Industry. (Annual)
   Data on production, sales, markets, capacities of the plastics industry.

492. *Handbook of Thermoset Plastics.* Edited by Sidney H. Goodman. Park Ridge, NJ: Noyes Data Corp., 1986. 421 p.
   Current data, descriptions, technology, and applications of thermoset plastics.

493. *Industrial Synthetic Resins Handbook.* Edited by Ernest W. Flick. Park Ridge, NJ: Noyes Data Corp, 1985. 388 p.
   Data on almost 3,000 current resins and related products with a section on suppliers' addresses.

494. *Modern Plastics Encyclopedia, 1986/87.* New York: McGraw-Hill, Inc. 878 p. (Annual)
   General information on plastics materials and processes; lists of suppliers.

495. *Packaging Reference Issue, Including the 1986 Encyclopedia.* Vol. 31. Newton, MA: Cahners Magazine Division, Reed Holdings, Inc. (Annual)
   Materials forecast, statistics of packaging, salary surveys, and a year-end index.

496. *Plastic Films for Packaging: Technology, Applications and Process Economics.* By Calvin J. Benning. Brookfield Center, CT: Society of Plastics Engineers, 1983. 192 p.
   A guide and reference to the materials, manufacturing methods, and markets of plastic packaging films today.

497. *U.S. Foamed Plastics Markets & Directory, 1987.* Brookfield Center, CT: Society of Plastics Engineers, 1985. 68 p.
   Lists companies by twenty-five product/service classifications. Also provides a review and projections of various end-use markets for different foam plastics.

498. *Whittington's Dictionary of Plastics.* 2d ed. By Lloyd R. Whittington. Brookfield Center, CT: Society of Plastics Engineers, 1978. 344 p.

Indexes, Abstracts, and Continuing Services

499. *Current Packaging Abstracts.* Piscataway, NJ: Rutgers, the State University, Gottscho Packaging Information Center. 1969–. (Semimonthly)

500. *International Packaging Abstracts.* Elmsford, NY: Pergamon Press, Inc. 1944–. (Monthly)

501. *Trends in End Use Markets for Plastics.* Enfield, CT: Springborn Laboratories, Inc. 1968–. (Monthly)
   Abstracts gleaned from foreign and domestic publications on new products, applications, end uses, and current events in the plastics industry.

Newspapers, Journals, and Newsletters

502. *Aerosol Age.* Cedar Grove, NJ: Industry Publications, Inc. (Monthly)
   Publishes the Buyers Guide Issue in October.

503. *Crittenden Plastic and Rubber Buyers.* Novato, CA: Crittenden News Service, Inc. (Weekly)

504. *Food and Drug Packaging.* New York: Magazines for Industry, Inc. (Monthly)

505. *Modern Plastics.* New York: McGraw-Hill, Inc. (Monthly)
   Publishes the Materials and Markets Review in January featuring statistics on resin sales, market growth, and end use patterns.

506. *Packaging.* Newton, MA: Cahners Publishing Co. (Monthly)
   Special issues include:
   March—Packaging Encyclopedia
   October—Buyers Guide

507. *Packaging Digest.* Chicago: Delta Communications, Inc. (Monthly)

508. *Plastics Brief.* Toledo, OH: Market Search, Inc. (Weekly)

509. *Plastics Design Forum.* Denver, CO: Industry Media, Inc. (Bimonthly)

510. *Plastics Engineering.* Brookfield Center, CT: Society of Plastics Engineers, Inc. (Monthly)

511. *Plastics Focus.* New York: Plastics Focus Publishing Co., Inc. (Weekly)

512. *Plastics Industry News.* Denver, CO: HBJ Publications. (6/year)

513. *Plastics Technology.* New York: Bill Communications. (Monthly)
   Special issues include:
   January—Manufacturing Outlook
   April—Injection Molding Survey

514. *Plastics World.* Newton, MA: Cahners Publishing, Division of Reed Publishing. (Monthly)
Features a regular section on resin prices. Special issues include:
January—Top 500 Plastics Processing Plants 50 Outstanding Materials of the Year
March—Plastics World Directory

515. *Statistical Report on Thermoplastic and Thermosetting Resins.* New York: Society of the Plastics Industry; Available from Ernst & Whinney, Trade Association Services Department. (Monthly)

516. *Urethane Plastics and Products.* Lancaster, PA: Technomic Publishing Co. (Monthly)

U.S. Government Publications

517. U.S. Department of Commerce. Office of Business Analysis. *The U.S. Plastics and Synthetic Materials Industry Since 1958.* Washington, DC: Government Printing Office, 1985. 122 p.

RUBBER

Motor vehicle tires are a large percentage of the product shipments in this industry. Fabricated rubber products are also a part of this category.

Electronic Retrieval Systems

518. RAPRA ABSTRACTS. Shrewsbury, England: Rubber and Plastics Research Association of Great Britain.
Offers commercial and technical information pertinent to the rubber and plastics industries. Also available on BRS and PERGAMON INFOLINE.

519. WARD'S AUTOINFOBANK. Detroit, MI.
A database with information on vehicle deliveries, production, inventories, body styles, and other data for the U.S. and Canada.

Encyclopedias, Dictionaries, and Handbooks

520. *Automotive News Market Data Book Issue.* Chicago: Crain Communications, 1987. 208 p. (Annual)
Statistics on production, sales, suppliers, prices, products, specifications, manufacturers, dealers, equipment, and other information.

521. *MVMA Facts & Figures, 1986.* Detroit, MI: Motor Vehicle Manufacturers Association, 1986. 96 p. (Annual)
Numbers, charts, graphs, and editorial content on motor vehicle products.

522. *Rubber Directory and Buyers Guide (RUBBICANA), 1987.* Akron, OH: Crain Communications, Inc. 800 p. (Annual)

523. *Rubber Red Book: A Comprehensive Directory of Manufacturers and Suppliers to the Rubber Industry.* 10th ed. Atlanta, GA: Communications Channels, Inc., 1987. 600 p. (Annual)
   Lists rubber goods manufacturers, services, equipment, and distributors.

524. *Ward's Automotive Yearbook.* 49th ed. Detroit, MI: Ward's Communications, Inc., 1987. 264 p. (Annual)
   Gives a variety of industry statistics; lists automotive suppliers and product guide to parts suppliers.

Indexes, Abstracts, and Continuing Services

525. *Current Literature in Traffic and Transportation.* Evanston, IL: Northwestern University Transportation Center. (Bimonthly)
   Distributed by the Transportation Library.

Newspapers, Journals, and Newsletters

526. *Automotive Industries.* Radnor, PA: Chilton Co. (Monthly)
   Special issues include:
   April—Specifications and Statistics Issue
   December—Auto Industry Report

527. *Automotive News.* Chicago: Crain Communications. (Weekly)
   Special issues include:
   April—Market Data Book
   December—Forecast

528. *Elastomerics.* Atlanta, GA: Communication Channels, Inc. (Monthly)

529. *MEMA Market Analysis.* Teaneck, NJ: Motor & Equipment Manufacturers Association. (Bimonthly)

530. *Rubber & Plastics News.* Akron, OH: Crain Automotive Group, Inc., Crain Communications, Inc. (Fortnightly)
   The April and October issues feature information on urethanes.

531. *Rubber Manufacturers Association Statistical Report. Industry Rubber Report.* Washington, DC: Rubber Manufacturers Association. (Monthly)

532. *Rubber World.* Akron, OH: Lippincott & Peto. (Monthly)
   The October issue focuses on chemicals and materials.

533. *Tire Review.* Akron, OH: Babcox Publications, Inc. (Monthly)

534. *Ward's Auto World.* Detroit, MI: Ward's Communications, Inc. (Monthly)

535. *Ward's Automotive Reports.* Detroit, MI: Ward's Communication, Inc. (Weekly)

U.S. Government Publications

536. U.S. Department of Transportation. Federal Highway Administration. *Traffic Volume Trends.* Washington, DC: Government Printing Office. (Monthly)

TEXTILES

This industry category focuses on the output from weaving and finishing mills, yarn mills, knitting mills, and carpet and rug mills. Natural and man-made fibers are also a part of this group.

Encyclopedias, Dictionaries, and Handbooks

537. *Encyclopedia of Textiles, Fibers, and Nonwoven Fabrics.* Edited by Martin Grayson. 1984. 581 p.

538. *Fairchild's Textile and Apparel Financial Directory.* 13th ed. New York: Fairchild Publications, Inc., 1986. 150 p.
   General industry trends and product profiles. Data on publicly owned textile and apparel corporations.

Indexes, Abstracts, and Continuing Services

539. *Man-Made Fiber Producers Handbook.* Roseland, NJ: Textile Economics Bureau, Inc. (Monthly, Quarterly, Semiannually, Annually)
   Provides information on capacities, inventories, imports, exports, end uses, and shipments by fiber type.

540. *Textile Technology Digest.* Charlottesville, VA: Institute of Textile Technology, 1944–. (Monthly)
   International coverage of the literature on textiles and related subjects.

541. *World Textile Abstracts.* Manchester, England: Shirley Institute, 1969–. (Semimonthly)
   Worldwide literature on all aspects of the textile industry, including end uses, production, international trade, and technical economics. Also available on PERGAMON INFOLINE.

Newspapers, Journals, and Newsletters

542. *America's Textiles International.* Atlanta, GA: Billiam Publishing Co. (Monthly)
   Special issues include:
   January—Outlook
   June—Financial Issue

543. *Apparel World.* New York: National Knitwear & Sportswear Association. (Monthly)

544. *Carpet & Rug Industry.* Ramsey, NJ: Rodman Publications, Inc. (Monthly)
   Special issues include:
   June—Top 25 Carpet Mills
   November—Dye and Chemical Buyers Guide

545. *DNR (Daily News Record).* New York: Fairchild Publications. (Daily)

546. *International Man-Made Fibre Production Statistics.* Zurich, Switzerland: International Textile Manufacturers Association. (Quarterly)

547. *Knitting Times.* New York: National Knitwear & Sportswear Association. (Weekly)

548. *Man-Made Fiber Review.* Roseland, NJ: Textile Economics Bureau. (Monthly)

549. *Nonwovens Industry.* Ramsey, NJ: Rodman Publications, Inc. (Monthly)
   Special issues include:
   January—Diaper Update
   November—Fibers/Chemicals/Films Directory
   December—Year-end Wrap-up Forecast Issue.

550. *Textile Business Outlook.* Norwich, NY: Statistikon Corp. (3/year)

551. *Textile Hi-Lights.* Washington, DC: American Textile Manufacturers Institute. (Quarterly)

552. *Textile Industries.* Atlanta, GA: W.R.C. Smith Publishing Co. (Monthly)
   The January or February issue presents the carpet industry forecast, and industry trends.

553. *Textile Organon.* Roseland, NJ: Textile Economics Bureau, Inc. (Monthly)

554. *Textile Pricing Outlook.* Norwich, NY: Statistikon Corp. (3/year)

555. *Textile World.* Atlanta, GA: McGraw-Hill Publications Co. (Monthly)
   The August issue presents a carpet forecast.

556. *Women's Wear Daily.* New York: Fairchild Publications. (Daily)

U.S. Government Publications

557. U.S. Congress. Senate. Committee on Finance. Subcommittee on International Trade. *State of the U.S. Textile Industry: Hearing.* Washington, DC: Government Printing Office, 1984. 334 p.

Associations

Many organizations and associations publish large amounts of information of an economic nature that are pertinent to the chemical industry.

GENERAL

American Association for the Advancement of Science. 1333 H St., NW, Washington, DC 20005.

American Chemical Society. 1155 16th St., NW, Washington, DC 20036.
   An organization composed of people involved in all branches of chemistry. They provide educational activities, legislative monitoring, publications, and other services. There are 32 divisions and most have regular meetings and publish papers.

American Institute of Chemical Engineers. 345 E. 47th St., New York, NY 10017.
   This professional group establishes standards for chemical engineering curricula and also collects statistics on chemical engineers.

The American Institute of Chemists, Inc. 7315 Wisconsin Ave., Bethesda, MD 20814.
   Founded to serve the personal and professional interests of chemists and chemical engineers.

Chemical Manufacturers Association. 2501 M St., NW, Washington, DC 20037.
   Manufacturers of basic chemicals who sell a substantial portion of their production to others.

Chemical Marketing Research Association. 139 Chestnut Ave., Staten Island, NY 10305.
   Composed of persons engaged in chemical market research for chemical companies in such fields as rubber, glass, soap, textiles, plastics, steel, metals, petroleum, drugs, and food.

Chemical Specialties Manufacturers Association. 1001 Connecticut Ave., Washington, DC 20036.
   Manufacturers, marketers, formulators, and suppliers of household, industrial, and personal care chemical specialty products.

National Association of Purchasing Management. 496 Kinderkamack Rd., Oradell, NJ 07649.
Composed of purchasing managers of a variety of firms.

National Science Foundation. 1800 G St., NW, Washington, DC 20550.
An independent agency of the executive branch of the federal government concerned with basic and applied research and education in the sciences.

Synthetic Organic Chemical Manufacturers Association. 1330 Connecticut Ave., Scarsdale, NY 20036.
Manufacturers of products derived from coal, natural gas, crude petroleum, and other natural substances and their derivatives.

AGRICULTURAL CHEMICALS, FOOD AND BEVERAGES

Distilled Spirits Council of the United States. Suite 900, 1250 Eye St., Washington, DC 20005.
Producers and marketers of beverage and distilled spirits sold in the U.S.

The Fertilizer Institute. 1015 18th St., NW, 11th Fl., Washington, DC 20036.
Producers, manufacturers, importers, brokers, and dealers of fertilizer and fertilizer materials and equipment.

Food Marketing Institute. Suite 700, 1750 K St., NW, Washington, DC 20006.
Grocery retailers and wholesalers.

Institute of Food Technologists. 221 N. LaSalle St., Chicago, IL 60601.
Technical personnel in food industries; production product development and research.

International Foodservice Manufacturers. 875 N. Michigan Ave., Chicago, IL 60611.
National and international manufacturers and processors of food, food equipment, and related products for the away-from-home food market.

National Agricultural Chemicals Association. 1155 15th St., NW, Suite 900, Washington, DC 20005.
Members are companies who produce or formulate agricultural chemical products.

National Fertilizer Solutions Association. 10777 Sunset Office Dr., Suite 10, St. Louis, MO 63127.
Manufacturers, wholesalers, and dealers of nitrogen solutions, fertilizers, chemicals, and additives.

National Food Processors Association. 1401 New York Ave., NW, Washington, DC 20005.
Commercial packers of fresh, frozen, or preserved food products.

National Soft Drink Association. 1101 16th St., NW, Washington, DC 20036.
Manufacturers of soft drinks.

CLEANING PREPARATIONS, COSMETICS AND TOILETRIES

Cosmetic, Toiletry and Fragrance Association. 1110 Vermont Ave., NW, Washington, DC 20005.
Manufacturers and distributors of finished cosmetics, fragrances, and toilet preparations; also suppliers of raw materials and services.

Fragrance Foundation. 142 E. 30th St., New York, NY 10016.
Members are perfume manufacturers, publishers, advertising agencies, suppliers, etc.

Fragrance Materials Association of the United States. 900 17th St., NW, Suite 658, Washington, DC 20006.
Manufacturers of fragrance and fragrance ingredients.

Independent Cosmetic Manufacturers & Distributors, Inc. Box 727, Bensenville, IL 60106.
Members are the small cosmetic manufacturer distributor and retailer.

International Sanitary Supply Association, Inc. 5330 N. Elston Ave., Chicago, IL 60630.
Manufacturers and distributors of cleaning and maintenance products.

Soap & Detergent Association. 475 Park Ave., South, New York, NY 10016.
Manufacturers of soap, detergents, fatty acids, and glycerine.

Society of Cosmetic Chemists. 1995 Broadway, 17th Fl., New York, NY 10023.
Members are scientists who are involved in the cosmetics industry.

DRUGS, PHARMACEUTICALS

American Hospital Association. 840 N. Lake Shore Dr., Chicago, IL 60611.
Individuals and healthcare institutions.

American Pharmaceutical Association. 2215 Constitution Ave., NW, Washington, DC 20037.
Professional society of pharmacists.

American Society of Hospital Pharmacists. 4630 Montgomery Ave., Bethesda, MD 20814.
Pharmacists employed by hospitals and related institutions.

Drug, Chemical and Allied Trades Association. 42–40 Bell Blvd., Suite 604, Bayside, NY 11361.
Manufacturers of drugs, chemicals, and related products.

Pharmaceutical Manufacturers Association. 1100 15th St., NW, Suite 900, Washington, DC 20005.
Manufacturers of ethical pharmaceutical and biological products.

United States Pharmacopeial Convention. 12601 Twinbrook Parkway, Rockville, MD 20852.
Composed of authorities in medicine, pharmacy, and the allied sciences.

ELECTRONICS, ELECTRICAL

American Electronics Association. 2670 Hanover, P.O. Box 10045, Palo Alto, CA 94303.
Manufacturers of electronics components and equipment.

American Public Power Association. 2301 M St., NW, Washington, DC 20037.
Members are publicly owned electric utility systems.

Association of Home Appliance Manufacturers. 20 N. Wacker Dr., Chicago, IL 60606.
Members are companies who manufacture most of the appliances sold in the U.S.

Edison Electric Institute. 1111 19th St., NW, Washington, DC 20036-3691.
Investor-owned electric utility companies operating in the U.S.

Electrochemical Society. 10 S. Main St., Pennington, NJ 08534-2896.
Technical society of electrochemists, chemists, and others interested in electrochemistry, electronics, and other subjects.

Electronic Industries Association. 2001 Eye St., NW, Washington, DC 20006.
Manufacturers of electronic parts, tubes, and solid state components; radio, television, and video systems; audio equipment, government electronic systems, and industrial and communications electronic products.

Institute of Electrical and Electronics Engineers. 345 E. 47th St., New York, NY 10017.
Engineers and scientists in electrical engineering, electronics, and allied fields.

National Electrical Manufacturers Association. 2101 L St., NW, Washington, DC 20037.
Manufacturers of electrical power equipment.

North American Electric Reliability Council. Research Park, Terhune Rd., Princeton, NJ 08540.
Regional organizations of electric utilities which represent systems operating in all the states and parts of Canada. They are concerned with bulk electric power supply.

Semiconductor Equipment and Materials Institute. 625 Ellis St., Suite 212, Mountain View, CA 94043.
Membership includes corporations or individuals and supply fabrication equipment, materials, or services to the semiconductor industry.

Semiconductor Industry Association. 4320 Stevens Creek Blvd., Suite 275, San Jose, CA 95129.
Producers of semiconductor products such as discrete components, integrated circuits, and microprocessors.

MINING, MINERALS, METALS

American Bureau of Metal Statistics. Box 1405, Plaza Station, Secaucus, NJ 07094.
Nonferrous metal producers.

American Ceramic Society. 65 Ceramic Dr., Columbus, OH 43214.
Scientists, engineers, and others involved in glass, ceramics, refractories, electronics, and structural clay products industries.

American Concrete Institute. Box 19150, Redford Station, Detroit, MI 48219.
Engineers, architects, contractors, and others involved in the technical aspects of concrete.

American Institute of Mining, Metallurgical and Petroleum Engineers. 345 E. 47th St., New York, NY 10017.
Professionals in the fields of mining, metallurgical, and petroleum engineering.

American Iron and Steel Institute. 1000 16th St., NW, Washington, DC 20036.
Basic manufacturers and individuals in the steel industry.

American Mining Congress. 1920 N St., Suite 300, NW, Washington, DC 20036.
Domestic producers of coal, metals, and minerals.

American Nuclear Society. 555 N. Kensington Ave., LaGrange Park, IL 60525.
Members are chemists, physicists, or others with professional experience in nuclear science.

American Society for Metals. Metals Park, OH 44073.
Metallurgists and executives in metals producing and consuming industries.

Compressed Gas Association. 1235 Jeff Davis Highway, Arlington, VA 22202.
Companies engaged in producing and distributing compressed, liquefied, and cryogenic gases. They also manufacture related equipment.

National Coal Association. 1130 17th St., NW, Washington, DC 20036.
Producers and sellers of coal and equipment.

National Stone Association. 1130 17th St., NW, Washington, DC 20036.
Producers and processors of crushed stone used for construction, chemical, metallurgical, and agricultural processes.

Portland Cement Association. 5420 Old Orchard Rd., Skokie, IL 60077.
Manufacturers of portland cement in the U.S. and Canada.

PAINTS, VARNISHES, LACQUERS, ENAMELS, COATINGS, ADHESIVES AND SEALANTS

Adhesive and Sealant Council. 1500 N. Wilson Blvd., Suite 515, Arlington, VA 22209.
Firms who manufacture and sell all rubber and plastic-based adhesives and related sealants.

Adhesives Manufacturers Association. 111 E. Wacker Dr., Chicago, IL 60601.
Manufacturers of paper converting and packaging adhesives, glue, gum, hot melts, and paste.

Chemical Coaters Association. Box 241, Wheaton, IL 60189.
Users and suppliers of industrial cleaners, paints, coatings, and equipment.

Color Association of the U.S. 343 Lexington Ave., New York, NY 10016.
Members are interested in colors used in various industries such as dyestuffs, paints, textiles, etc.

Federation of Societies for Coatings Technology. 1315 Walnut St., Philadelphia, PA 19107.
Promotes research on paints, varnishes, coatings, pigments, and other related materials.

Gummed Industries Association. 380 N. Broadway, Jericho, NY 11753.
Members are converters of gummed paper products and suppliers of basic raw materials.

National Coil Coaters Association. 1900 Arch St., Philadelphia, PA 19103.
Manufacturers of coated metal coil and suppliers of materials used in coil coatings.

National Paint and Coatings Association. 1500 Rhode Island Ave., NW, Washington, DC 20005.
Members are manufacturers of paints and chemical coatings and suppliers of raw materials and equipment.

PAPER AND ALLIED PRODUCTS

American Forest Institute. 1619 Massachusetts Ave., NW, Washington, DC 20036.

American Paper Institute. 260 Madison Ave., New York, NY 10016.
U.S. manufacturers and converters of pulp and paper, wood products, landowners, and wood preservers.

Forest Products Research Society. 2801 Marshall Ct., Madison, WI 53705.
Individuals interested in the wood industry; research, production, utilization.

Hardwood Plywood Manufacturers Association. Box 2789, Reston, VA 22090-2789.
Manufacturers and prefinishers of hardwood plywood; also suppliers of glue, veneer, and other products to the industry.

Institute of Paper Chemistry. 1043 E. South River St., Appleton, WI 54912.
Manufacturers of pulp, paper, or paperboard.

National Association of Home Builders. 15th and M Sts., NW, Washington, DC 20005.

National Forest Products Association. 1619 Massachusetts Ave., NW Washington, DC 20036.
A federation of forest products associations and companies.

Pulp Chemicals Association. 60 E. 42d St., Rm. 824, New York, NY 10165.
Manufacturers of chemical products produced from wood pulp.

Technical Association of the Pulp and Paper Industry. Box 105113, Technology Park, Atlanta, GA 30348.
Executives, managers, engineers, research scientists, and technologists in the pulp, paper, and allied industries.

PETROCHEMICALS, ENERGY

American Gas Association. Arlington, VA.
Distributors and transporters of natural, manufactured, and liquefied gas.

American Petroleum Institute. 1220 L St., NW, Washington, DC 20005.
Producers, refiners, marketers, and transporters of petroleum and allied products including crude oil, lubricating oil, gasoline, and natural gas.

International Association of Drilling Contractors. Box 4287, Houston, TX 77210.
Oilwell contract drilling firms.

National LP-Gas Association. 1301 W. 22d St., Oak Brook, IL 60521.
Producers and distributors of butane-propane gas; also manufacturers and distributors of equipment and utilization appliances.

PLASTICS MATERIALS, PACKAGING

Can Manufacturers Institute. 1625 Massachusetts Ave., NW, Washington, DC 20036.
Suppliers of cans.

Chemical Fabrics & Film Association. 1230 Keith Bldg., Cleveland, OH 44115-2180.
Manufacturers of vinyl and urethane products.

Closure Manufacturers Association. 6845 Elm St., Suite 208, Mclean, VA 22101.
Manufacturers of metal and plastic closures.

Composite Can and Tube Institute. 1742 N St., NW, Washington, DC 20036.
Manufacturers of composite cans, tubes, spools, cores, mailing packages, etc.

Flexible Packaging Association. 1090 Vermont Ave., NW, Suite 500, Washington, DC 20005.
Converters of paper, foils, and plastics packaging materials.

National Association of Plastics Distributors. 5001 College Blvd., Leawood, KS 66211.
Distributors of plastics materials, piping, and resins to the industry.

National Paperbox and Packaging Association. 231 Kings Highway East, Haddonfield, NJ 08035.
Materials suppliers and manufacturers of rigid and folding paperboxes.

Packaging Institute, U.S.A. 20 Summer St., Stamford, CT 06901.
Manufacturers of packaging materials, machinery, and services.

Plastic Shipping Container Institute. 150 N. Wacker Dr., Suite 1145, Chicago, IL 60606.
High Density Polyethelyne (HOPE) producers and container manufacturers.

Polyurethane Manufacturers Association. 800 Roosevelt Rd., Suite C 20, Glen Ellyn, IL 60137.

Society of Plastics Engineers. 14 Fairfield Dr., Brookfield Center, CT 06805.
An international technical and professional society dedicated to promoting scientific and engineering knowledge relative to plastics.

Society of the Plastics Industry. 355 Lexington Ave., New York, NY 10017.
Manufacturers and processors of molded, extruded, fabricated, laminated, and reinforced plastics.

RUBBER

Energy Rubber Group, Inc. 910 Cypress Station Rd., Houston, TX 77090.
Concerned with elastomeric products used in the energy field.

International Institute of Synthetic Rubber Producers. 2077 S. Gessmer, Suite 133, Houston, TX 77063.
Synthetic rubber manufacturers.

Malaysian Rubber Bureau (U.S.A.). 1925 K St., NW, Washington, DC 20006.
Provides service to rubber consumers and manufacturers.

Motor and Equipment Manufacturers Association. 222 Cedar Lane, Teaneck, NJ 07666.

Motor Vehicle Manufacturers Association of the U.S., Inc. 300 New Center Bldg., Detroit, MI 48202.
Association members are domestic car, truck, and bus manufacturers.

Natural Rubber Shippers Association, Inc. 1400 K St., NW, 9th Fl., Washington, DC 20005.

Rubber Manufacturers Association. 1400 K St., NW, Washington, DC 20005.
Manufacturers of tires, tubes, mechanical and industrial products, footwear, sporting goods, and other rubber products.

Society of Automotive Engineers, Inc. 400 Commonwealth Dr., Warrendale, PA 15096.

Tire & Rim Association, Inc. 3200 W. Market St., Akron, OH 44313.

TEXTILES

American Association of Textile Chemists and Colorists. Box 12215, Research Triangle Park, NC 27709.
Technical and scientific society of textile chemists and colorists in textile and related industries using colorants and chemical finishes.

American Textile Manufacturers Institute. 1101 Connecticut Ave., NW, Suite 300, Washington, DC 20036.
Textile mill firms who manufacture and process cotton, manmade wool, and silk textile products.

Industrial Fabrics Association International. 345 Cedar Bldg., Suite 450, St. Paul, MN 55101.
Fiber producers, weavers, coaters, laminators, finishers, and producers and manufacturers of canvas and industrial fabric end products and their suppliers.

Man-Made Fiber Producers Association. 1150 17th St., NW, Washington, DC 20036.
Manufacturers of man-made fibers used in apparel, household goods, industrial materials, and other products.

National Knitwear & Sportswear Association. 386 Park Ave., South, New York, NY 10016.
Manufacturers of a variety of knitted wearing apparel.

Consultant Sources

Sources used to find experts in specific fields are listed as well as individual consultants specializing in offering services to the chemical industry.

DIRECTORIES

Bradford's Directory of Marketing Research Agencies and Management Consultants in the United States and the World. Fairfax, VA: Bradford's Directory, 1986. 288 p.
    Lists firms that provide services in the marketing area, including management consultants, testing laboratories, etc.

Consultants and Consulting Organizations Directory. 7th ed. Detroit, MI: Gale Research Co., 1986. 1,750 p. 2 vols.
    A reference guide to concerns and individuals engaged in consultation for business, industry, and government.

Consulting Services. 23d ed. New York: Association of Consulting Chemists and Chemical Engineers, 1986.
    Lists members of the association who are qualified chemists or chemical engineers presently engaged in consulting practice. Published biannually.

Directory of Chemical Engineering Consultants. 6th ed. Edited by Mary Pat Healy. New York, NY: American Institute of Chemical Engineers, 1986. 49 p.
    Lists full-time and part-time consultants and other experts in areas specific to the chemical process industry.

Directory of U.S. and Canadian Marketing Surveys and Services. 6th ed. Bridgewater, NJ: Rauch Associates, Inc., 1986. 430 p.
    Lists multiclient marketing reports and continuing services available for 300 consulting firms.

FINDEX; The Directory of Market Research Reports, Studies & Surveys. 8th ed. New York: Find/SVP, 1986. Annual.
    A directory of market research reports, studies, and surveys. It contains descriptions of consumer and industrial studies and surveys, syndicated and multiclient studies, audits and subscription research services, as well as published reports on general management and business topics.

Greenbook; International Directory of Marketing Research Houses and Services. 24th ed. New York: American Marketing Association, 1986. 460 p.

GENERAL

Business Communications Co., Inc. 9 Viaduct Rd., P.O. Box 2070C, Stamford, CT 06906.
A business research firm that provides market and technical research skills covering a broad range of popular subjects. They specialize in reports that forecast trends in various businesses.

Business Trend Analysts. 2171 Jericho Turnpike, Commack, NY 11725.
A market research firm covering popular subjects in diversified industries. Their expertise is in providing market research skills (mainly focus groups and telephone interviews).

Catalytica Associates, Inc. 430 Ferguson Dr., Mountain View, CA 94043.
The firm concentrates on the technical and economic aspects of the catalysts field; providing development and market studies.

Chase Econometrics. 1414 Bond Court Building, Cleveland, OH 44114.
Provides market evaluations relative to current and anticipated economic demands. Economic forecasts and market studies concerning existing and new products.

Ducker Research Co. 4050 W. Maple Rd., Birmingham, MI 48010.
Economic and technical aspects of the construction industry. Market studies, product development, and opportunity analysis.

Frost & Sullivan, Inc. 106 Fulton St. New York, NY 10038.
Covers a range of subjects in diversified industries. Specialty is in analyzing and forecasting market trends.

Gorham International, Inc. P.O. Box 8, Gorham, ME 04038.
Technical and business development in numerous industries. Specializes in high-tech issues.

Hochberg & Co., Inc. Chester Professional Building, Chester, NJ 07930.
Business aspects of the chemical industries. Specializes in mining, chemicals, textiles.

Hull & Co. 5 Oak St., P.O. Box 4250, Greenwich, CT 06830.
Business and technical aspects of chemical industries.

International Research Associates. 2041 Whitehorse Rd., Berwyn, PA 19312.
Focuses on technical and economic aspects of specialty chemicals.

International Trader Publications. 144 Hunter Ave., North Tarryton, NY 10591.
This firm publishes monthly trade reports on various materials, concentrating on basic chemicals and plastics.

Charles Kline & Co. 330 Passaic Ave., Fairfield, NJ 07006.
Specializes in chemical marketing research and reports concentrating on historic data.

Omega Research Associates, Inc. 5741 Centre Ave., Pittsburgh, PA 15206.
Business and technical skills in varied chemical industries through literature searching and interviews with key government and industry contacts.

Oxenham Technology Associates, Inc. 310 E. 46 St., Suite 16L, New York, NY 10017.
Expertise in catalysts, process materials, petrochemicals.

Predicasts, Inc. 200 University Circle Research Center, 11001 Cedar Ave., Cleveland, OH 44106.
A research firm that provides business and marketing information in the form of periodicals, reports, and multiclient and custom studies.

Probe Economics. 105 Kisco Ave., New York, NY 10549.
Specializes in market studies with an extensive computerized database and statistical modeling for forecasting in various industry segments.

Puckorius & Associates, Inc. 1202 Highway 74, Suite 211, P.O. Box 2440, Evergreen, CO 80439.
Information on water treatment industries.

Quantum Enterprises, Inc. P.O. Box 307, Lincroft, NV 07738.
Construction industries.

Schotland Business Research, Inc. 19 Van Kerk Rd., Princeton, NJ 08540.
Polymers and packaging.

Peter Sherwood Associates, Inc. 20 Haarlem Ave., White Plains, NY 10603.
Petrochemical and thermoplastic industries.

Strategic Analysis, Inc. Box 3485, R.D., 3, Fairlane Rd., Reading, PA 19606.
Specializes in acquisitions and strategic planning for specialty chemical and electronics industries.

Technomic Consultants. 100 Corporate North, Bannockburen, IL 60015.
Packaging, ceramics, diversified.

Tecnon, Ltd. 11 Hay Hill, London W1X 7LF England.
Provides marketing surveys and economic analyses of the chemical industry.

Phillip Townsend Associates, Inc. P.O. Box 90327, Houston, TX 77290.
Market research and analysis in petrochemicals, plastics, paper, forest products, metals, and glass industries.

Wharton Econometric Forecasting Associates. 3624 Market St., Philadelphia, PA 19104.
Provide forecasting and analytical services.

AGRICULTURAL CHEMICALS, FOOD AND BEVERAGES

Chase Econometrics Associates, Inc. 150 Monument Rd., Bala Cynwd, PA 19004.
   Agricultural Service. Analyzes and forecasts short- and long-term prices and quantities of all major U.S. livestocks and crops.
Delphi Marketing Services, Inc. 400 E. 89th St., New York, NY 10028.
   Provides market research in the technical and economic aspects of the food industry.
Doane Marketing Research, Inc. 555 N. New Ballas Rd., St. Louis, MO 63141.
   Provides marketing research, management information, and consultation to agribusiness firms.
Jack Frost and Associates. P.O. Box 6429, Lawrenceville, NJ 08648.
   A market research firm concentrating on the economic business and technical issues of food processing.
The Hereld Organization. 401 Christopher Ave., Suite 11, Gaithersburg, MD 20877.
   Consultants to suppliers of the food and beverage industries.
Nuventures Consultants, Inc. P.O. Box 2489, LaJolla, CA 92038.
   Proprietary and multiclient studies on chemicals.
Strategy Development Consultants, Inc. 1600 Broadway, Suite 1950, Denver, CO 80202.
   This firm specializes in the technical and economic aspects of various industries, with special expertise in food and agriculture.

CLEANING PREPARATIONS, COSMETICS AND TOILETRIES

C.A. Houston & Associates, Inc. 434 Mamaroneck, Mamaroneck, NY 10545.
   Expertise in surfactants and related areas. Market studies with technical and business support.

DRUGS, PHARMACEUTICALS

Biomedical Business International. 17722 Irvine Blvd., Tustin, CA 92680.
   Covers the field of medical devices and diagnostics.
Robert S. First, Inc. 707 Westchester Ave., White Plains, NY 10604.
   Business and management aspects of medical and pharmaceutical fields. Extensive surveys, literature searches, and interviews.

ELECTRONICS, ELECTRICAL

Electronic Trend Publications. 10080 N. Wolfe Rd., Suite 372, Cupertino, CA 95014.
Provides studies and market research in data processing and electronic markets.

Electronicast. Bay View Tower, 2121 South El Camino Real, Suite 1215, San Mateo, CA 94403.
Specialize in forecasting technology and markets for the electronics industries.

Gnostic Concepts, Inc. 951 Mariner's Island Blvd., Suite 300, San Mateo, CA 94404 (Division of McGraw-Hill).
Multiclient and proprietary studies on the electronics and electrical industries. Also provide strategic planning and acquisition analysis services.

Integrated Circuit Engineering. 15022 N. 75th St., Scottsdale, AZ 85260.
Area of concentration is the production of technical and technoeconomic studies involving electronics and related industries.

International Planning Information, Inc. 164 Pecora Way, Pecora Valley, CA 94025.
Reports cover the electronics and information industries; market research and business strategies.

Kessler Marketing Intelligence. 22 Farewell St., Newport, RI 02840.
Offers economic- and business-oriented skills and information in the fiber optics industry.

Reach Associates, Inc. 75 S. Orange Ave., South Orange, NJ 07079.
Performs technoeconomic research on a multiclient and private basis in several areas including electronic chemicals.

Rose Associates. 111 Main St., Los Altos, CA 94022.
Concentrates on technical and economic aspects in the electronics field.

MINING, MINERALS, METALS

Charles River Associates, Inc. John Hancock Tower, 200 Claredon St. Boston, MA 02116.
Concentration of expertise is consultant services in mineral, metals, and material areas.

CRU Consultants, Inc. 33 W. 54th St., New York, NY 10019.
Provide economic and market analysis of metals in industries.

Fallon Research Associates, Inc. P. O. Box 944, New Providence, NJ 07974.
Market research in specialty minerals, metals, and other raw materials.

Roskill Information Services, Ltd. 2 Clapham Rd., London SW9 OJA England.
Worldwide research reports on metals and minerals.

PAINTS, VARNISHES, LACQUERS, ENAMELS, COATINGS, ADHESIVES AND SEALANTS

Chemark Information Services. 9916 Carver Rd., P.O. Box 43214, Cincinnati, OH 45242.
Provide business and technological information with particular strengths in adhesives and coatings.

H. S. Holappa & Associates. 9 Hart Rd., Box 243, Lynnfield, MA 01940.
Experience and expertise in the technical and business aspects of adhesives.

Weh Corp. P.O. Box 40066, San Francisco, CA 94140.
Specializes in corporate planning, acquisitions, and mergers for formulated chemical industries. Expertise in coatings and resins.

PAPER AND ALLIED PRODUCTS

Andover International Associates. 130 Centre St., Danvers, MA 01923.
Provides economic and technical research skills to management consulting for the pulp and paper industry.

Jaakko Poyry International OY. P.O. Box 16, SP. 00441, Helsinki 40 Finland.
Consulting and engineering and market research for the forest industries.

PETROCHEMICALS, ENERGY

Chemical Data, Inc. 830 Brookhollow Central, 2900 N. Loop West, Houston, TX 77092.
Provide multiclient and custom studies in the petroleum and petrochemical industries.

Chemical Market Associates, Inc. 11757 Katy Freeway, Suite 750, Houston, TX 77079.
A market research firm that addresses the technical and economic aspects of the petrochemical industry on a global basis.

DeWitt & Co. 3650 Dresser Tower, 601 Jefferson St., Houston, TX 77002.
This firm offers technical and business expertise in petrochemical industries.

Pace Co. Consultants and Engineers, Inc. 5251 Westheimer, P.O. Box 53473, Houston, TX 77052.
Specializes in economic analysis and forecasting for hydrocarbon industries.

PLASTIC MATERIALS, PACKAGING

George E. Ham. 284 Pine Rd., Briarcliff Manor, NY 10510.
Business and technical research in the polymer industries.

R. M. Kossoff & Associates, Inc. P.O. Box 389, Milburn, NJ 07041.
Economic and technical focus on thermoplastics; analyses of competitive materials, suppliers, and uses of plastics in various industrial segments.

Margolis Marketing and Research Co. 232 Madison Ave., New York, NY 10016.
Specialize in providing technical and market information in the plastics industry.

Market Search, Inc. P.O. Box 2886, Toledo, OH 43606.
Technical and economic information in the plastics industry.

Monkman-Rumsey. P.O. Box 3760, Wilmington, DE 19807.
Management and technical aspects of specialty chemical industries. Specialize in high-tech applications, end-uses, and competitive assessment in chemical markets.

MSK Associates. 1648 Stonybrook, Rochester, MI 48063.
Offers technical and business skills in the chemical industry. Specialize in military topics (particularly aerospace engineering).

Sabre Associates, Inc. P.O. Box 486, Belden Station, Norwalk, CT 06852.
Plastics, packaging, and specialty chemicals.

Schotland Business Research, Inc. P.O. Box 511, Princeton, NJ 08542.
Polymers and packaging.

Skeist Laboratories, Inc. 112 Naylon Ave., Livingston, NJ 07039.
Polymers, engineering plastics.

Springborn, Group, Inc. One Springborn Center, Enfield, CT 06082.
Raw materials in plastics industries.

TPC Business Group. 851 Holland Ave., P.O. Box 3535, Lancaster, PA 17604.
Plastics and polymers.

TEXTILES

Geoffrey Lund Associates. 335 Rocky Rapids Rd., Stamford, CT 06903.
Market research for the textiles and fibers industry.

Kurt Salmon Associates, Inc. 400 Colony Square, 9th Fl., Atlanta, GA 30361.
A consumer product-oriented consultancy with expertise in carpet and textile markets.

RBI International Carpet Consultants. P.O. Box 722, Dalton, GA 30720.
Concentration in this consultancy is on the carpet industries.

75 Statistikon Corp. P.O. Box 246, East Norwich, NY 11732.
International management consultants in textiles, economics, and engineering.

Author Index

Numbers in italic refer to page numbers. All other numbers refer to citation numbers.

Antony, Arthur, 1
Arthur D. Little Decision
  Resources, 84
Ash, Irene, 74, 502
Ash, Michael, 74, 502

Battelle, 85
Bennett, H., 45, 51
Benning, Calvin J., 496, 509

Chem Systems, Inc, 87
Clark, Ronald L., 58
The Conference Board, 89

Daniells, Lorna M., 7
Dickson, Cheryl L., 458-60
Dun's Marketing Services, 90

Evinger, William R., 54

Faith, W. L., 58
Flick, Ernest W., 506

Gardner, William, 57
Goodman, Sidney H., 505
Grant, Claire, 56
Grant, Roger, 56
Grayson, M., *2*, 549
Griffiths, Mary C., 262
Gruber, Katherine, 50

Hampel, Clifford A., 52, 55
Hawley, Gessner G., 46, 52, 55
Healy, Mary Pat, *73*
Heilberger, C. A., 503

Kent, James A., 71
Keyes, Donald B., 58
Kuney, Joseph H., 38

Landrock, Arthur H., 395
Langenkamp, Robert D., 450
Lavigne, John R., 451
Lewis, Richard J., 46

Maizell, R. E., 3
Ministry of Chemical Industry
  (Beijing, China), 42

Nass, L. I., 503

Parker, Sybil P., 64-65
Peck, Theodore P., 4

Quadra Associates, 3

Ramachandran, V. S., 341

Sainer, E., 251
Sax, N. Irving, 46
Shelton, Ella Mae, 458-59
SRI, International, 94
Standard & Poor's Corp., 95
Sturchio, Jeffrey L., 5

Thackray, A., 40

U.S. Department of Agriculture.
  Crop Reporting Board, 212

U.S. Department of Agriculture.
    Crop Reporting Board.
    Statistical Reporting Service,
    213
U.S. Department of Agriculture.
    Economic Research Service,
    214-16
U.S. Department of Agriculture.
    Economics Management Staff,
    217
U.S. Department of Agriculture.
    Economics Management Staff.
    Information Division, 218-20
U.S. Department of Agriculture.
    Foreign Agricultural Service.
    Information Division, 221-22
U.S. Department of Agriculture.
    Statistical Reporting Service,
    223
U.S. Department of Commerce, 140
U.S. Department of Commerce.
    Bureau of Economic Analysis,
    141
U.S. Department of Commerce.
    Bureau of Industrial
    Economics, 422
U.S. Department of Commerce.
    Bureau of the Census, 142-47,
    224, 370
U.S. Department of Commerce.
    Industry and Trade
    Administration. Office of
    Field Operations, 148
U.S. Department of Commerce.
    International Trade
    Administration, 271
U.S. Department of Commerce.
    International Trade Com-
    mission, 149
U.S. Department of Commerce.
    International Trade Admin-
    istration. Industry Analysis
    Division, 272
U.S. Department of Commerce.
    Office of Business Analysis,
    517
U.S. Department of Energy. Energy
    Information Administration,
    323, 371-73, 475-80
U.S. Department of Energy. Energy
    Information Administration.
    Office of Coal, Nuclear,
    Electric, and Alternate Fuels,
    324-25
U.S. Department of Energy. Energy
    Information Administration.
    Office of Energy Markets and
    End Use, 374, 481-82
U.S. Department of Energy. Energy
    Information Administration.
    Office of Oil & Gas, 483
U.S. Department of Energy. Office
    of Utility Project Operations,
    326
U.S. Department of the Interior.
    Bureau of Mines, 375-80
U.S. Department of Transportation.
    Federal Highway
    Administration, 536
U.S. Executive Office of the
    President. Council of
    Economic Advisers, 150
U.S. Federal Reserve System. Board
    of Governors, 151
U.S. House of Representatives.
    Committee on Energy and
    Commerce, 484
U.S. International Tariff
    Commission, 152
U.S. International Trade
    Commission, 153-55
U.S. Office of the Federal Register,
    156-58

Whittington, Lloyd R., 498
Windholz, Martha, 258
Wolman, Yecheskel, 6
Woodward, Paul W., 459-60

Title Index

Numbers in italic refer to page numbers. All other numbers refer to citation numbers.

Abstract Bulletin of the Institute of Paper Chemistry, 404
ACS Directory of Graduate Research, 34
Adhesives Age, 387
Adhesives Age Directory, 381
Adhesives and Sealants Newsletter, 388
Adhesives Technology Handbook, 382
Advertising Age, 96
Aerosol Age, 502
Agmarketer, 181
AGRIBUSINESS U.S.A, 160
Agrichemical Age, 182
AGRICOLA, 161
Agricultural Economics Research, 218
Agricultural Exports Outlook and Situation, 214
Agricultural Outlook, 219
Agricultural Prices, 213
Agricultural Statistics, 223
Agriculture and Food. Abstract Newsletter, 176
AGRIDATA NETWORK, 162
AGRINDEX, 177
Agri Marketing, 183
Agri Marketing; Marketing Services Guide Issue, 165
Air Conditioning, Heating & Refrigeration News, 287
Aldrich Catalog/Handbook of Fine Chemicals, 35

The Almanac of the Canning, Freezing, Preserving Industries, 166
American Druggist, 253
American Electronics Association; Membership Directory, 1987, 273
American Glass Review, 344
American Hospital Association Guide to the Health Care Field, 239
American Journal of Agricultural Economics, 184
American Journal of Agricultural Economics Handbook Directory, 159
American Laundry Digest, 227
American Men and Women of Science; Physical and Biological Sciences, 8
American Metal Market, 345
American Paint & Coatings Journal, 389
America's Textiles International, 542
Annual Bulletin of Trade in Chemical Products, 1984, 36
Annual Outlook for United States Electric Power, 324
Annual Review of the Chemical Industry, 1984, 37
APILIT, 428
Apparel World, 543
Appliance, 288
Appliance Manufacturer, 289

Automotive Industries, 526
Automotive News, 527
Automotive News Market Data Book Issue, 520

Basic Petroleum Data Book, 1987, 433
Beverage Industry, 185
Beverage Industry Annual Manual, 1987, 167
Beverage World, 186
Beverage World Data Bank, 1986/87, 168
Biomedical Products, 254
BIOSCAN; The Biotechnology Corporate Directory Service, 251
Biotechnology News, 187
Boxboard Containers, 408
Bradford's Directory of Marketing Research Agencies and Management Consultants in the United States and the World, 73
Brick and Clay Record, 346
BRS/SEARCH SERVICE, 17
The B-Tip Program, Battelle Technical Inputs to Planning, 85
Building Supply and Home Centers, 409
Business America, 140
Business Information Sources, 7
Business Intelligence Program, 94
Business Marketing, 97
Business Periodicals Index, 75
Business Week, 98

CAPPS-Chemicals and Polymer Production Statistics, 86
Carpet & Rug Industry, 544
CEH ON-LINE, 18
Cellular Business, 290
Census of Agriculture, 1982, 224
Census of Manufactures, 1982, 142
Census of Mineral Industries, 1982, 370
Ceramic Abstracts, 340
Ceramic Industry, 347
C4 Market Report, 450
C4 Monitor, 451
Chemcyclopedia 87, 38
Chemical Business, 100

Chemical Economics Handbook, 94
Chemical Economy and Engineering Review (CEER), 101
Chemical Engineering, 102
Chemical Engineering Faculties, 1986-87, 9
Chemical & Engineering News, 99
Chemical Industries Information Sources. Management Information Guide Series, 4
Chemical Industries Newsletter, 103
Chemical Industry Growth in Developing Countries and Changing U.S. Trade Patterns, 153
Chemical Industry Notes, 76
Chemical Industry Update: North America, 104
Chemical Information: A Practical Guide to Utilization, 6
Chemical Insight, 105
Chemical Marketing & Management, 106
Chemical Marketing Reporter, 107
Chemical Profiles, 88
Chemical Spotlight, 108
Chemical Times & Trends, 109
Chemical Week, 110
Chemical Week Buyers Guide, 1986, 39
Chemistry and Industry, 111
Chemistry in America, 1876-1976: Historical Indicators, 40
Chemscope, 112
Chemsources-U.S.A, 41
Chem Systems, Inc., 88
CHEMTECH, 113
Chemweek Newswire, 114
China Chemical Industry 1985/86; World Chemical Industry Yearbook, 42
Cleaning Management, 228
Coal Age, 348
Coal Outlook, 349
Coal Week, 350
The Code of Federal Regulations, 156
Coinamatic Age, 229
College Chemistry Faculties, 43
Commerce Business Daily, 148
Commercial Atlas & Marketing Guide, 44
Commodity Year Book, 1987, 169

Commodity Yearbook Statistical Abstract Service, 178
Communications Week, 291
A Competitive Assessment of the U.S. Pharmaceutical Industry, 272
Compilation of Selected Energy Related Legislation, 484
Compressed Air, 351
Concise Chemical and Technical Dictionary, 45
Concrete Admixtures Handbook; Properties, Science and Technology, 328
Concrete Products, 352
Condensed Chemical Dictionary, 46
Construction Review, 422
Consultants and Consulting Organizations Directory, 73
Consulting Services, 73
Consumer Electronics, 292
Corporate History and the Chemical Industries; A Resource Guide, 5
Cosmetic World News, 230
Cosmetics & Toiletries, 231
County and City Data Book; A Statistical Abstract Supplement, 143
CPI Digest, 77
CPI Purchasing, 115
CRIS, 163
Crittenden Plastic and Rubber Buyers, 503
Crop Production, 212
Crow's Buyers and Sellers Guide of the Forest Products Industries, 1986, 396
CURRENT BIOTECHNOLOGY ABSTRACTS, 164
Current Industrial Reports, 144
Current Literature in Traffic and Transportation, 525
Current Packaging Abstracts, 499

Data Communications, 293
DATA RESOURCES, INC, 19
Dealerscope Merchandising, 294
Defense Electronics, 295
Desk & Derrick Standard Oil Abbreviator, 434

DIALOG INFORMATION RETRIEVAL SERVICES, Inc., 20
Dictionary of Biotechnology, 240
Diesel Fuel Oils, 1985, 445
Directory of American Research and Technology, 47
Directory of Chemical Engineering Consultants, 73
Directory of Chemical Producers, U.S. & Western Europe, 94
Directory of Custom Chemical Manufacturers, 48
Directory of Iron and Steel Works of the U.S. and Canada, 329
The Directory of the Canning, Freezing, Preserving Industries, 170
Directory of the Forest Products Industry, 1986, 397
Directory of U.S. and Canadian Marketing Surveys and Services, 73
Directory of World Chemical Producers, 49
DNR (Daily News Record), 545
Domestic Uranium Mining and Milling Industry (1985 Viability Assessment), 371
DOW JONES NEWS RETRIEVAL, 21
Drug and Cosmetic Catalog, 1987, 241
Drug and Cosmetic Industry, 232, 255
Drugstore News, 256
Drug Topics, 257
Drug Topics Red Book, 1987, 242
Dun's Business Month, 116

The Economic Bulletin Board, 22
Economic Indicators, 150
EE—Electronic/Electrical Product News, 296
Elastomerics, 528
Electric Light & Power, 297
Electric Utility Week, 298
Electrical & Electronics Abstracts, 286
Electrical World, 299
Electronic Business, 300
Electronic Business Forecast, 301

Electronic Business 200 Issue, 1986, 275
Electronic Chemicals & Materials News, 302
Electronic Chemicals News, 303
Electronic Market Data Book, 276
Electronic Market Trends, 304
Electronic Marketing Directory, 1986, 277
Electronic Materials Report, 305
Electronic News, 306
Electronic News Financial Fact Book & Directory, 1986, 278
Electronic Packaging & Production, 307
Electronic Products Magazine, 308
Electronic Reporting System for the Monthly Statistical Report—RESINS, 486
Electronics, 309
Electronics Buyers Guide. 1986/87, 279
ELSS—Electronic Legislative Search System, 23
Encyclopedia of Associations, 50
Encyclopedia of Chemical Trademarks and Synonyms, 51
The Encyclopedia of Chemistry, 52
Encyclopedia of Electronics, 280
Encyclopedia of Packaging Technology, 488
Encyclopedia of Plastics, Polymers & Resins, 489
Encyclopedia of PVC, 490
Encyclopedia of Textiles, Fibers, and Nonwoven Fabrics, 537
Engineering & Mining Journal, 353
Engineering and Mining Journal International Directory of Mining and Mineral Processing Operations, 1986, 330
Engineering News Record, 354
EPUB Electronic Publication System, 429
Ethylene Service Newsletter, 452
Europa Yearbook, 53
European Chemical News, 117

Facts & Figures of the U.S. Plastics Industry, 1986, 491
Fairchild's Textile and Apparel Financial Directory, 538

Farm Chemicals, 188
Farm Chemicals Handbook, 1987, 171
Farm Journal, 189
Farmline, 221
Farm Supplier, 190
FAST TRACK, 327
The Federal Register, 157
Federal Reserve Bulletin, 151
Federal Statistical Directory, 54
Federation of Societies for Coatings Technology—Yearbook and Membership Directory, 383
Feedstuffs, 191
Feedstuffs—Reference Issue, 1986, 172
Fiber Intermediates Market Report, 453
Fiber Optics News, 310
Fiberoptics Report, 311
The Fiber Optics Sourcebook, 281
Financial Statistics of Selected Electric Utilities, 1984, 325
Financial Times International Yearbook—Mining, 331
Financial Times Who's Who in World Oil and Gas, 1982-83, 423
FINDEX; The Directory of Market Research Reports, Studies & Surveys, 73
Food & Beverage Marketing, 192
Food and Drug Packaging, 504
Food Chemical News, 193
Food Development, 194
Food Engineering, 195
Food Ingredients Directory, 173
Foodlines, 196
Food Processing, 197
Food Processing Ingredients, Equipment & Supplies Guide & Directory, 174
Food Science and Technology Abstracts, 179
Food Technology, 198
Foods Adlibra, 180
Forbes, 118
The Forefront Biotechnology Service, 84
Foreign Agricultural Trade of the United States, 220
Foreign Agriculture, 222
Forest Industries, 410

Title Index 87

Fortune, 119
Frozen Food Digest, 199
Future Markets and Petroleum Supply, 481

Gas Facts, 1985, 435
Genetic Engineering and Biotechnology Related Firms Worldwide Directory, 1987/88, 243
Glass Factory Directory, 1986, 332
Glass Industry, 355
The Glass Industry Directory, 1987, 333
Glossary of Chemical Terms, 55
Grant and Hackh's Chemical Dictionary, 56
Greenbook; International Directory of Marketing Research House and Services, 74
Green Markets, 200
The Green Sheet; Weekly Pharmacy Reports, 258
Guide to Basic Information Sources in Chemistry. Information Resources Series., 1
Gulf Coast Oil Directory, 1987, 436

Handbook of Basic Economic Statistics, 120
Handbook of Chemical Synonyms and Trade Names, 57
Handbook of Thermoset Plastics, 492
Health Care Industry Service, 84
Health Industry Today, 259
Heating Oils, 1986, 446
High Technology, 312
Historical Statistics of the United States Colonial Times to 1970, 145
Household & Personal Products Industry, 233
How to Find Chemical Information; A Guide for Practicing Chemists, Teachers & Students, 3
Hybrid Circuit Technology, 313
Hydrocarbon Processing, 454

Illustrated Petroleum Reference Dictionary, 437
Industrial Chemical News, 121

Industrial Chemicals, 58
Industrial Finishing, 390
Industrial Minerals, 356
Industrial Minerals and Rocks, 334
Industrial Minerals Directory; World Guide to Producers and Processors, 1986, 335
Industrial Synthetic Resins Handbook, 493
Industry Surveys, 95
Industry Week, 122
Information Retrieval in Chemistry and Chemical Patent Law, 2
InfoTran, 84
Inside FERC'S Gas Market Report, 455
Integrated Information Systems Service, 84
International Energy Outlook, 1985: With Projections to 1995, 482
International Fiber Optics and Communications; Handbook and Buyer's Guide Issue, 1987, 282
International Financial Statistics, 123
International Man-Made Fibre Production Statistics, 546
International Packaging Abstracts, 500
International Petroleum Encyclopedia, 1986, 438
International Petroleum Finance, 456
International Pharmaceutical Abstracts, 252
International Who's Who in Energy and Nuclear Sciences, 424
The International Who's Who of the Arab World, 425
Inventory of Power Plants in the United States, 1985, 326
INVESTEXT, 24
Iron Age, 357
Iron Ore Databook, 336

Japan Chemical Directory, 59
Japan Chemical Week, 124
Japan Company Handbook, 60
JCW Chemicals Guide, 1986/87, 61
Journal of Coatings Technology, 391
Journal of Commerce and Commercial, 125

Key Chemicals & Polymers, 62
Keystone Coal Industry Manual, 1986, 337
Kiplinger Agricultural Letter, 201
Kirk-Othmer Encyclopedia of Chemical Technology, 63
Knitting Times, 547

Laser Report, 314
Lasers & Applications, 315
Lawn and Garden Marketing, 202
Lawn Care Industry, 203
Lockwood's Directory of the Paper and Allied Trades, 1987, 398

Main Economic Indicators, 126
Man-Made Fiber Producers Handbook, 539
Man-Made Fiber Review, 548
Mannsville Chemical Product Synopses, 91
Manufacturing Chemist (formerly Chemical Age)., 127
Marketing News, 128
Marketing Times, 129
Marketletter, 260
Marquis Who's Who in America, 1986-87, 10
McCutcheon's Emulsifiers and Detergents, North American Edition, 225
McGraw-Hill Concise Encyclopedia of Science & Technology, 64
McGraw-Hill Dictionary of Chemical Terms, 65
Medical and Healthcare Marketplace Guide, 244
Medical Device & Diagnostic Industry, 261
Medical Devices, Diagnostics and Instrumentation Reports: The Gray Sheet, 262
Medical Equipment and Supplies Worldwide, 271
Medical Marketing & Media, 263
Medical World News, 264
MEMA Market Analysis, 529
The Merck Index; An Encyclopedia of Chemicals, Drugs, and Biologicals, 245
Metal Bulletin, 358
Metal Statistics, 1986, 338
Metals Abstracts, 341

Metals Week, 359
Metals Week Insider Report (Telex Service), 360
The Million Dollar Directory Series, 90
Mineral Commodity Summaries, 1987, 375
Mineral Facts and Problems, 376
Mineral Industry Surveys, 377
The Mineral Position of the United States: The Past Fifteen Years, 1971-1985, 378
Mineral Producers and Processors Directory, 1985, 339
Mineralogical Abstracts, 342
Minerals Yearbook, 379
Mining Magazine, 361
Modern Brewery Age, 204
Modern Healthcare, 265
Modern Metals, 362
Modern Paint & Coatings, 392
Modern Plastics, 505
Modern Plastics Encyclopedia, 1986/87, 494
Monomers Market Report, 457
Monthly Bulletin of Statistics, 130
Monthly Energy Review, 374
Monthly Report on Pulp and Paper Mill Projects in the World, 411
Monthly Statistical Bulletin, 458
Monthly Statistical Summary, 412
Moody's Manuals and News Reports, 92
Motor Gasolines, Summer, 1985, 447
MVMA Facts & Figures, 1986, 521

National Association of Chain Drug Stores—Membership Directory, 246
National Food Review, 217
National Petroleum News, 459
NATIONAL REGISTRY OF WOMEN IN SCIENCE AND ENGINEERING, 11
National Trade and Professional Associations of the United States, 66
National Wholesale Druggist's Association—Membership and Executive Directory, 1987, 247
Natural Gas Annual, 1986, 475
Natural Gas Liquids Update, 460

Natural Gas Monthly, 476
NERAC, 25
New Scientist, 131
NEWSNET, 26
NEXIS, 27
The 1985 Annual Report of the Secretary of the Interior Describing the Nonfuel Mineral Industry; Supply, Demand and Outlook, 380
Nonwovens Industry, 549

OECD Observer, 132
OIL AND GAS JOURNAL ENERGY DATABASE, 430
Oil & Gas Journal, 461
Oil & Gas Journal Data Book, 1986, 439
Oil and Gas Stocks Handbook, 440
Oil Industry Comparative Appraisals, 462
Oil Industry Outlook, 1987–1991, 441
Oil Industry, U.S.A, 442
Olefins Market Letter, 463
OPD Chemical Buyers Directory, 67
Opportunities in Chemical Distribution; Dynamics of a Growing Industry, 68

Packaging, 506
Packaging Digest, 507
Packaging Reference Issue, Including the 1986 Encyclopedia, 495
Paint and Coatings Industry, 393
Paint Red Book, 384
Painting and Wallcovering Contractor—PDCA Yearbook Issue, 385
Paperboard Packaging, 416
Paper, Film & Foil Converter, 413
Paper Industry Management Association—Membership Directory, 1986/87, 395
Paper Industry News Digest, 405
Paper, Paperboard & Wood Pulp Capacity, 414
Paper Trade Journal, 415
Paper Year Book, 1987, 399
Performance Chemicals, 133
PERGAMON INFOLINE, 28
Pest Control, 205

Petrochemical News, 464
Petroleum Abstracts, 448
Petroleum Economist, 465
PETROLEUM/ENERGY BUSINESS NEWS, 431
Petroleum/Energy Business News Index, 449
Petroleum Marketing Annual 1985, 477, 478
Petroleum Marketing Monthly, 483
Petroleum Supply Monthly, 479
Pharmacy Times, 266
Physicians' Desk Reference, 1986, 248
PIERS, 29
The Pink Sheet, 268
Pipeline & Gas Journal, 467
Pipeline Digest Who's Who in Pipelining, 1983, 426
Pipe Line Industry, 466
PIRA Abstracts, 406
Pit & Quarry, 363
PLASPEC, 487
Plastic Films for Packaging: Technology, Applications and Process Economics. By, 496
Plastics Brief, 508
Plastics Design Forum, 509
Plastics Engineering, 510
Plastics Focus, 511
Plastics Industry News, 512
Plastics Technology, 513
Plastics World, 514
Platt's Oilgram & Price Report, 468
PMA Newsletter, 267
Polymers/Ceramics/Composites Alert, 78
Post's Pulp & Paper Directory, 1986, 400
Power, 316
Prepared Foods, 206
Principle International Business, 90
Printed Circuit Fabrication, 317
Process Economics Program, 94
Process Evaluation and Research Planning Service (PERP), 87
Product Marketing and Cosmetic and Fragrance Retailing, 235
Products Finishing Magazine, 394
Profiles of Eminent American Chemists, 12
Profiles of U.S. Chemical Distributors, 69

PROMT, 79
Propylene Service Newsletter, 469
PTS Funk & Scott Indexes, 80
Public Power, 318
Public Utilities Fortnightly, 319
Pulp & Paper, 417
Pulp & Paper Capacities Survey, 1984-1989, 401
Pulp and Paper Dictionary, 402
Pulp & Paper International Newswire, 407
Pulp & Paper—North American Industry Factbook, 1984-85, 403
Pulp & Paper Week, 418
Purchasing World, 134

Quick Frozen Foods, 207
Quick Frozen Foods' Directory of Frozen Food Processors and Buyers Guide, 1987, 175

Random Lengths, 419
RAPRA ABSTRACTS, 518
Recent Additions to Baker Library, 81
Register of Corporations, Directors and Executives, 95
The Report on Performance Materials, 135
Research Centers Directory 1987, 70
Riegel's Handbook of Industrial Chemistry, 71
Rock Products, 364
The Rome Report, 93
The Rose Sheet, 234
Rubber & Plastics News, 530
Rubber Directory and Buyers Guide (RUBBICANA), 1987, 522
Rubber Manufacturers Association Statistical Report. Industry Rubber Report, 531
Rubber Red Book: A Comprehensive Directory of Manufacturers and Suppliers to the Rubber Industry, 523
Rubber World, 532

S and MM Sales & Marketing Management, 136
Sanitary Maintenance, 236
Scientific Meetings, 137

SCRIP—World Pharmaceutical New, 269
SDC/ORBIT Search Service, 30
The Shift from U.S. Production of Commodity Petrochemicals to Value Added Specialty Chemical Products and the Possible Impact on U.S. Trade, 154
SIA Circuit Newsletter, 320
Skillings Mining Review, 365
Snack Food, 208
Soap/Cosmetics/Chemical Specialties, 237
Soap/Cosmetics/Chemical Specialties Blue Book, 1986, 226
Solid State Technology, 321
Specialty Chemicals Handbook, 72
Specialty Chemicals Update Program, 94
Standard & Poor's Register of Corporations, Directors and Executives, 13
State and Metropolitan Area Data Book; A Statistical Abstract Supplement, 146
Statistical Abstract of the United States, 1987, 147
Statistical Panorama, 1987, 283
Statistical Report on Thermoplastic and Thermosetting Resins, 515
Statistical Service, 95
Statistical Yearbook, 1983/84, 73
Statistical Yearbook of the Electric Utility Industry, 1987, 284
Steam-Electric Plant Construction Cost and Annual Production Expenses, 323
Survey of Current Business, 141

Tariff Schedules of the United States Annotated, 1987, 152
Technology Update, 82
Telephony, 322
Textile Business Outlook, 550
Textile Hi-Lights, 551
Textile Industries, 552
Textile Organon, 553
Textile Pricing Outlook, 554
Textile Technology Digest, 540
Textile World, 555
Thesaurus of Chemical Products, 74

Tire Review, 533
Toluene-Xylenes Newsletter, 470
TRADSTAT, 31
Traffic Volume Trends, 536
Trends in End Use Markets for Plastics, 501
TULSA, 432

The United States Government Manual, 1986/87, 158
Uranium Industry Annual 1985, 372
Urethane Plastics and Products, 516
U.S.A. Oil Industry Directory, 443
USAN and the USP Dictionary of Drug Names, 249
U.S. Foamed Plastics Markets & Directory, 1987, 497
U.S. Glass, Metal & Glazing, 367
U.S. Industrial Outlook, 1987, 149
U.S. Oil Week, 471
The U.S. Plastics and Synthetic Materials Industry Since 1958, 517
U.S. Production and Sales of Synthetic Organic Chemicals, 1985, 155

VU-TEXT, 32

Wall Street Journal, 138
Wall Street Transcript, 139
WARD'S AUTOINFOBANK, 519
Ward's Automotive Reports, 535
Ward's Automotive Yearbook, 524
Ward's Auto World, 534
Washington Drug Letter, 270
Weed, Trees and Turf, 209
Weekly Coal Production, 373
Weekly Petroleum Status Report, 480
Weekly Propane Newsletter, 472
Weekly Statistical Bulletin, 473

What's New in Advertising and Marketing, 83
Whittington's Dictionary of Plastics, 498
Who Owns Whom, 90
Who's Who in Electronics, 1987, 274
Who's Who in Finance and Industry, 25th ed, 14
Who's Who in Frontiers of Science & Technology, 15
Who's Who in Health Care, 238
Who's Who in Packaging, 485
Who's Who in Technology, 16
Who's Who in World Petrochemicals, 427
WILSONLINE, 33
Women's Wear Daily, 556
Wood Products, 420
World Agricultural Supply and Demand Estimates, 215
World Agriculture: Outlook and Situation, 216
World Aluminum Abstracts, 343
World Coal, 368
World Directory of Pharmaceutical Manufacturers, 250
World Electronics Yearbook, 1987, 285
World Food & Drink Report, 210
World Mining, 369
World Oil, 474
World Surface Coatings, 386
The World Telecommunications Service, 84
World Textile Abstracts, 541
Worldwide Petrochemical Directory, 444
World Wood, 421

Yard & Garden, 211
33 Metal Producing, 366

Subject Index

Numbers in italic refer to page numbers. All other numbers refer to citation numbers.

Adhesives, 77, 110, 382-83, 388-89
  associations, *69*
  consultants, *78*
Advertising, 83, 94, 96
Aerosols, 503
Agricultural chemicals, 162-66, 172, 183, 189, 191, 201, 203-04, 210
  associations, *65*
  consultants, *76*
  marketing, 161-63, 166, 182, 184
Agriculture, 79, 165-66, 177, 185, 190, 202-15, 225
  crops, 213
  prices, 214
  statistics, 224-25
Air conditioning, 284, 288
Aluminum, 344
  foil, 414
Apparel, 539, 544, 546, 557
  associations, *72*
Appliances, 289-90, 293, 295
  associations, *67*
Associations, *64-72*
  directories, *50, 66*
Automotive, 520-22, 525, 527-28, 530, 535-37
  associations, *71-72*

Beverages, 168-69, 186-87, 193, 205, 211
  associations, *65*
  statistics, 205
Biographical directories, 8-16

Biotechnology, 165-88, 241, 252, 255
  companies, 244-45, 252
  dictionaries, 241
Brick, 347

Carpet, 545
  consultants, *80*
Cement, 329, 353, 364-65
  associations, *68*
Ceramics, 78, 136, 341, 348
  associations, *68*
Chemical industry, 37, 100, 103-04, 106, 108, 121, 127
  associations, *64-65*
  abstracts, 76-78
  Chinese, 42
  consultants, *73-76*
  history, 5, 40
  Japanese, 58, 60, 136
  statistics, 99, 105
Chemical engineers
  associations, *64*
Chemical information, 1-6, 63
Chemicals, 38, 40, 62-63, 67, 71-72, 74, 88-90, 94, 95, 99, 107, 246. *See also* Specialty chemicals
  dictionaries, 45-46, 55-57, 65
  distributors, 68-69
  electronic, 303-04, 306
  paper, 401, 405
  prices, 62, 88-89, 90, 92, 107, 114-15, 117, 125, 135
  producers, 38, 41, 48-49, 62, 67,

Subject Index 93

88-89, 90, 92, 99
  products, 38-39, 41, 48-49, 58, 62-63, 67, 71-72, 74, 77-78, 88-90, 92, 94, 95, 99, 107, 115
  rubber, 533
  shipping, 107, 125
  statistics, 40, 87-90, 92, 156
Chemistry, 52, 63, 71
Chemists, 8, 9, 11-12, 15-16
  associations, *64*
Clay, 347
Cleaning products, 226-30, 234, 237
  associations, *65*
  consultants, *76*
Coal, 349-51, 369, 364
  associations, *68*
  prices, 390, 393 See also Paint
Composites, 78, 136
Computers, 274-81
Concrete, 329, 353
  associations, *68*
Construction, 355, 423
Consultants, *73-80*
  directories, *73*
Contractors, 355
Corporations, 95
  international, 94
  Japanese, 59
  United States, 93-94, 116, 118-19, 139-40
Cosmetics, 107, 227, 231-33, 235-36, 238, 242, 256. See also Toiletries
  associations, *65*

Data communications, 292, 294, 323
Design firms, 355
Detergents, 107, 226, 238
  associations, *70*
Diagnostics, 262-63
Drilling, 431, 433, 449, 462, 475
Drugs, 79, 233, 242-43, 246, 249-50, 254, 256-59, 267, 269, 271. See also Pharmaceuticals
  associations, *66*
  companies, 247-49, 251
  consultants, *76*
  packaging, 505
Dry cleaning, 230

Elastomers, 529, 531, 533. See also Plastics and rubber
  associations, *71*
Electric power. See Utilities
Electronics, 79, 281, 286-87, 294, 296-97, 301-10, 311
  associations, *67*
  companies, 274-76, 278-80, 301
  consultants, *78*
  statistics, 277, 293
Energy, 79, 375, 483, 485
  consultants, *78*
Europe, 53
Exports. See Foreign trade

Fertilizer, 162, 183, 189, 191
  associations, *65*
  prices, 201
Fiberoptics, 282-83, 311-12
Fibers, 77, 136, 538, 540. See also Textiles
  associations, *72*
  consultants, *79*
Foam, 498 See also Plastics
Food, 79, 165-67, 171, 173-77, 179-81, 193, 12, 207-08, 218
  associations, *65*
  companies, 195-96, 198
  frozen, 200, 208
  ingredients, 196, 198, 209
  packaging, 209, 505
Foreign trade, 29, 31, 36, 153-55
Forest products, 397-98, 411, 420-22
  associations, *69, 70*
  consultants, *78*
Fragrances, 235-36
  associations, *66*
Fuel oil, 446

Gas, natural, 424, 427, 431, 433-34, 440-41, 455-56, 461-62, 468-77, 485
  associations, *70*
Gases, industrial, 352
  associations, *68*
Gasoline, 448
Genetic engineering, 244, 252
Glass, 333-34, 345, 356, 368
  companies, 333-34, 345
Government regulations, 157-59

Health care, 239-40, 245, 260-61, 266
  animal, 191
  associations, 66
Heating, 284, 288
  oil, 447
Herbicides, 191
Hospitals, 240, 253
  associations, 66

Imports. See Foreign trade
Inks, 77
Insecticides, 191, 210
Iron, 330, 337, 358
  associations, 68
  prices, 358

Lasers, 315-16
Laundry. See Detergents; Cleaning products
Lawn and garden, 191, 203-04, 210, 212
Legislation, 23. See also Government regulations
Lumber. See Forest products

Management, 84, 90
Maps, 44
Marketing, 79, 83, 89, 97, 128-29
Marketing research, 83-86, 88
  associations, 64
  consultants, 73
Medical, 267, 260-66, 272
  associations, 66
  consultants, 76-77
Metals, 136, 330-31, 339, 342, 346, 358-61, 363, 367-68, 380
  associations, 68
  consultants, 77
Minerals, 328, 331-32, 335-36, 343, 346, 354, 357, 376-81
  companies, 336, 340
  consultants, 77-78
  prices, 376
  statistics, 370-71, 376-81
Mining, 331-32, 362, 366, 370
  associations, 68
  consultants, 77

Packaging, 79, 407, 409, 414, 417, 486, 489, 496-97, 500-02, 505, 507-08
  associations, 71
  consultants, 79
Paint, 385-87, 390-95. See also Coatings
  associations, 69
  consultants, 78
  prices, 393
Paper, 79, 396, 399-402. See also Pulp
  associations, 69, 70
  consultants, 78
  statistics, 412-13, 415-19
Pesticides, 162, 206
Petrochemicals, 95, 107, 429, 445, 452, 455, 462, 464-65, 473
  consultants, 78
  prices, 451, 453, 454, 458, 470-71
Petroleum, 429, 442, 42, 433, 449-50, 457, 462, 466, 472, 475, 478-79
  biography, 424, 428
  companies, 437, 441, 463
  consultants, 78-79
  prices, 125, 469, 472, 475
  statistics, 430-35, 437-44, 459-60
Pharmaceuticals, 165, 253-54, 257, 259, 267-71, 273. See also Drugs
  associations, 66
  companies, 252
  consultants, 76-77
Pigments, 77. See also Paint
Plastics, 77, 79, 110, 486-88, 489-91, 493, 495, 497-99, 502, 504, 506, 509-11, 531
  associations, 70-71
  companies, 513, 515
  consultants, 79
  prices, 488, 506, 515
  statistics, 487, 492, 506, 516
Plywood. See Forest products
Polymers, 77-78, 87
  consultants, 79
Printed circuits, 314, 318, 321-22
Printing, 407
Pulp, 396, 399-404. See also Paper
  associations, 70
  consultants, 78
  statistics, 412-13, 415-19
Purchasing, 134-35
  associations, 64

Refrigeration, 284, 288
Research, 34, 43, 84-86, 99
 directories, 47, 70
Resins, 136, 494
Rock products, 335, 364-65
Rubber, 77, 504, 509, 523-24, 529, 531, 533
 associations, *71-72*
 statistics, 532
Rugs. *See* Carpet

Sales and selling, 137
Science and technology, 64, 82, 131
 associations, *64*
 meetings, 138
Semiconductors. *See* Electronics
Sealants, 368, 389. *See also* Adhesives
Soap. *See* Detergents
Specialty chemicals, 72, 94, 109, 133, 227, 238
 associations, *64-65*
 consultants, *79*
Statistics, 73, 95, 120, 123-24, 130, 144, 147
 historical, 146, 148, 151
 industry, 142-43, 145, 150
Steel, 330, 358

Stone. *See also* Rock products
 associations, *68*
Surface coatings. *See* Coatings and paint
Surfactants. *See* Cleaning products

Textiles, 538-44. *See also* Fibers
 associations, *72*
 consultants, *79-80*
Tires, 520, 534
 associations, *71-72*
Toiletries, 79, 232. *See also* Cosmetics
Trade names, 51, 57
Transportation, 526, 537

Uranium, 372-73
Urethanes, 517, 531
U.S. government, 54, 159
Utilities, 299-300, 317, 319-20, 324-25
 associations, *67-68*
 companies, 298, 326
 statistics, 285, 300, 326

Wallcoverings, 386
Weeds, 210